Control Engineering

Series Editor

Zongli Lin, Electrical and Computer Engineering, University of Virginia, Charlottesville, VA, USA

Advisory Editors

Amir Aghdam, Electrical & Computer Engineering, Concordia University, Montreal, QC, Canada

Frank Allgöwer, IST, Universität Stuttgart, Stuttgart, Germany

Alessandro Astolfi, Electrical & Electronic Engineering, Imperial College London, London, UK

Jay Farrell, Electrical & Computer Engineering, University of California, Riverside, Riverside, CA, USA

Jing Sun, Mechanical Engineering, University of Michigan, Ann Arbor, MI, USA

Ying Tan, Mechanical Engineering, University of Melbourne, Parkville, VIC, Australia

Gernot Herbst • Rafal Madonski

Active Disturbance Rejection Control

From Principles to Practice

Gernot Herbst
Electrical Engineering
University of Applied Sciences Zwickau
Zwickau, Germany

Rafal Madonski
Faculty of Automatic Control, Electronics
and Computer Science
Silesian University of Technology
Gliwice, Poland

ISSN 2373-7719   ISSN 2373-7727   (electronic)
Control Engineering
ISBN 978-3-031-72686-6   ISBN 978-3-031-72687-3   (eBook)
https://doi.org/10.1007/978-3-031-72687-3

This work was supported by Cleveland State University.

© The Editor(s) (if applicable) and The Author(s) 2025, corrected publication 2025. This book is an open access publication.

Open Access This book is licensed under the terms of the Creative Commons Attribution 4.0 International License (http://creativecommons.org/licenses/by/4.0/), which permits use, sharing, adaptation, distribution and reproduction in any medium or format, as long as you give appropriate credit to the original author(s) and the source, provide a link to the Creative Commons license and indicate if changes were made.
The images or other third party material in this book are included in the book's Creative Commons license, unless indicated otherwise in a credit line to the material. If material is not included in the book's Creative Commons license and your intended use is not permitted by statutory regulation or exceeds the permitted use, you will need to obtain permission directly from the copyright holder.
The use of general descriptive names, registered names, trademarks, service marks, etc. in this publication does not imply, even in the absence of a specific statement, that such names are exempt from the relevant protective laws and regulations and therefore free for general use.
The publisher, the authors and the editors are safe to assume that the advice and information in this book are believed to be true and accurate at the date of publication. Neither the publisher nor the authors or the editors give a warranty, expressed or implied, with respect to the material contained herein or for any errors or omissions that may have been made. The publisher remains neutral with regard to jurisdictional claims in published maps and institutional affiliations.

This book is published under the imprint Birkhäuser, www.birkhauser-science.com by the registered company Springer Nature Switzerland AG
The registered company address is: Gewerbestrasse 11, 6330 Cham, Switzerland

If disposing of this product, please recycle the paper.

Foreword

This book puts in the hands of practitioners, students, and scientists alike, a powerful tool known as Active Disturbance Rejection Control (ADRC). Both the operating principles and ensuing control laws are made fully understandable and easily executable. It addresses real-world control problems where complexity, uncertainty, and physical constraints abound, but solutions must remain simple and transparent.

Working at the forefront of technological developments for the past decade, the authors have played an instrumental role in shaping ADRC as a powerful platform for control system design in various sectors of engineering practice. It is with conviction that they set their minds on helping readers find a viable alternative to the century-old, but still dominating, trial-and-error practice of PID (proportional-integral-derivative) control, an alternative that is made both rational and practical in this book. The result is a well-structured book that brings the novices up the learning curve gently and gives the returning readers a source of references and a user's manual.

Chapter 2 begins with two physical examples and confronts the readers head-on with the central question: Exactly how much do we need to know about these dynamic systems in order to control them? Hence begins the "first contact with ADRC" and a rendition of ADRC as a paradigm shift in the field of automatic control. Here the style of the narrative is "show, don't tell," allowing readers with some background in college physics and math to follow the narrative in breaking the problems of control and arriving at the core ADRC design principles naturally.

Once a reader is engaged, there would be no turning back.

Chapter 2 is followed by concrete expositions in the subsequent chapters, as ADRC is generalized to N-th order systems in Chap. 3, interpreted and visualized in Chaps. 4 and 5, respectively, in both time and frequency domains, and extended and modified in Chap. 6 to demonstrate that "ADRC is not a fixed set of equations," but "a modular, flexible framework that could be tailored to specific applications."

Part I (Foundations) of the book ends in Chap. 7, which provides the overview of this active field of study and the bridge to Part II (Going Practical). Going practical, indeed! Covered in less than 50 pages, from Chaps. 8–10, are digitizing ADRC in both state space and transfer function forms, arranging bumpless transfer from one to another controller, dealing with physical imperfections such as actuator saturation, noise and deadtime, etc., and arriving finally at software implementation in Matlab/Simulink or C language directly. The key here is self-containment: practically it is all here for a user to try his/her hands on ADRC and solve real-world problems, as it is shown in Chap. 11.

In fact, the book doesn't have to be read sequentially. For example, readers may jump to Part II and do a quick proof of concept validation using the examples there, with the recipe-like procedures implemented in C codes or Simulink blocks. And then return to Part I for better understanding of the principles behind. Note that these procedures are collected as a whole in the final chapter for quick references, along with authors' outlook for ADRC as the tool of choice in the future.

Finally, I want to thank the authors personally for their dedication and unselfishness in writing this go-to book for a wide range of audiences. It is a distinct privilege of mine to witness their professional growth both as scholars and practitioners, a rare combination in itself, for more than a decade. They did a great service to the field of control engineering with this book by providing a pragmatic, self-contained introduction to ADRC, where principles and practices are intimately connected. They made every attempt to make the book reader-friendly, understandable by all, and plug-and-play ready. What they didn't cover in this compact book of fundamentals, such as the subject of rigorous mathematical analysis, they provided an overview of the landscape and references for further readings in Chap. 7.

Let the new era begin!

Cleveland, Ohio   Zhiqiang Gao
July 2024

Preface

Active disturbance rejection control (ADRC) has been the subject of active research for nearly two decades now. It witnessed many theoretical successes, including the derivation of rigorous stability proofs and the development of an understanding of various dynamical properties of active disturbance rejection schemes. During that time, several successful practical applications have been reported in this area. The last few years have seen impressive growth in the availability of commercial controllers based on the ADRC methodology.

We started working on ADRC in the early 2010s. Today, we have more than 20 years of combined experience dealing with both the theoretical and practical sides of ADRC. During this time, we have put out a significant body of work on that subject and our credentials in this area are widely available online. Through this book, we combine our knowledge gained in academia and industry to bring you a systematic and easily accessible way into the world of ADRC.

Why did we decide to write a book on ADRC, and why now? As the area of ADRC moves forward, we desire to provide a comprehensive and up-to-date introduction to it and make sure that the new generation has a solid foundation it can build on and improve upon. We intend to provide a book characterized by accessibility, clear language, and reproducibility of its contents. With the current surge of interest in ADRC from both academia and industry, it becomes very easy for newcomers to get overwhelmed by the number of new results on ADRC without a proper understanding of its basics. With this book, we want to offer a fresh presentation of the subject of ADRC, one that exploits its fundamentals while emphasizing its connection with classical control. Such treatment feels particularly timely, given the rapid advances in microprocessor and multi-processor technology which facilitate the utilization of advanced controllers. Limitations to the future growth of ADRC can thus hardly be expected to be computational, but rather from a lack of a fundamental understanding of methodologies for the design, evaluation, and testing of its algorithms. Therefore, we strive to make this book beneficial to people for whom it really is an introduction. In short: if you want to "talk" ADRC, you need to learn the language first.

This book is meant for a diverse audience. It is for all who share a common desire: to put things into practice! This uniting theme is important to us because we ourselves like to get things done so we wanted to write a book for people who want to do something with a practical impact. Something that they can take, simulate, and deploy. The book is constructed in a way that offers something for everyone, maybe except for people who expect theorem-proof style content. We offer diverse and easily accessible content that matches the subtitle of the book: "from principles to practice." Depending on the readers' backgrounds, we collect them along that way. For people starting from zero, we pick them up at the beginning. For those already knowledgeable, we pick them up on the way to practice. For the returning readers, we also have reference-style parts of the book that quickly guide to condensed formulae, ready to be used in software implementation. With all this in mind, the book can be used as a textbook, handbook, or reference manual, and be easily picked up by students, practitioners, and scientists. Whenever possible, we avoid jargon, and in the instances where we cannot, we break it down.

In the first chapter, we will explain our key motives that shaped the writing of this text. They heavily impacted its structure and led us to depart from some (unwritten) conventions. Whenever you, while reading, get a feeling this book might be different, these are the reasons—all for our goal: supporting you in putting ADRC into practice.

Acknowledgments

Gernot Herbst: Writing an introductory book on ADRC had long been on my mind. That it became *this* book was only possible by joining forces with Rafal Madonski. I could not have thought of a better coauthor for this undertaking! Special thanks go to Arne-Jens Hempel for many years of companionship on academic journeys, and Ringo Lehmann for selfless lab support (not only) for this book. For many fruitful discussions which sharpened my view on real-world control systems, I would like to thank my former colleagues of the power electronics R&D department at Siemens AG, Chemnitz, Germany, particularly Thomas Demuth, Frank Appelt, and Jörg Weiß. And, most importantly, everything would be nothing without my family.

Rafal Madonski: I would like to thank Professors Hebertt Sira-Ramírez, Shihua Li, Jun Yang, and my trusted coauthor Gernot Herbst for being a personal source of inspiration in the area of ADRC. I want to also thank Krzysztof Łakomy, Momir Stankovic, Mario Ramirez-Neria, Wenchao Xue, Sun Li, and Jacek Czeczot for their selfless support over the years. Finally, I dedicate my work in this book to my Family; to mother Barbara and her unwavering encouragement, to father Zbigniew who has been a constant, to sister Justyna and her moral fiber, and to wife Magda with whom we are never ready but this never stops us.

Preface

Both authors would like to express their heartfelt thanks to Professor Zhiqiang Gao. He not only sparked our interest to dive into the world of ADRC, but he remained a guiding light in the murky waters of the control field, and through most generous support, also made the open access publication of this book possible.

Zwickau, Germany
Poznan, Poland
July 2024

Gernot Herbst
Rafal Madonski

The original version of the book has been revised. A correction to this book can be found at https://doi.org/10.1007/978-3-031-72687-3

Contents

Part I Foundations

1 Prelude: A Fresh Look 3

2 First Contact with ADRC 11
   2.1 Control of First-Order Plants 11
   2.2 Towards Control of Higher-Order Systems 19

3 Linear Active Disturbance Rejection Control 29
   3.1 Derivation and Core Concepts 29
   3.2 Parameter Tuning 37
   3.3 Summary and Outlook 41

4 Between Time and Frequency Domains 45
   4.1 ADRC as a State-Space Controller 45
   4.2 Transfer Function Representation 49

5 Visual Tour 57
   5.1 Plant Modeling 57
   5.2 Closing the Loop: A Nominal Example 60
   5.3 The Role of ADRC's Tuning Parameters 62
   5.4 Coping with Parametric and Structural Plant Uncertainties 72

6 Extensions and Modifications 77
   6.1 Availability of Additional Model Information 77
   6.2 Availability of Reference Signal Derivatives 82
   6.3 Nonlinear ADRC 84
   6.4 Error-Based ADRC 91

7 Interlude: A Look Around 103

Part II Going Practical

8 Discrete-Time Linear ADRC 121
    8.1 State-Space Form 121
    8.2 Transfer Function Form 131
    8.3 Dual-Feedback Transfer Function Form 135
    8.4 Discrete-Time Error-Based ADRC 139
    8.5 Summary and Outlook 144
    References 146

9 Practical Aspects 149
    9.1 Controller Output Limitation 149
    9.2 Bumpless Transfer 153
    9.3 Dealing with Measurement Noise 157
    9.4 Control of Plants with Dead Time 159
    References 161

10 Software Implementation 163
    10.1 Implementation in MATLAB/Simulink 163
    10.2 Implementation in C Programming Language 165

11 Application Examples 171
    11.1 Heater Temperature Control 171
    11.2 DC-DC Converter Voltage Control 177

12 Postlude: A Look Ahead 183

Correction to: Active Disturbance Rejection Control C1

A Linear ADRC Cheat Sheet 191

B Overview of ADRC Implementations in Simulink 207

Index 213

Acronyms

ADRC	Active Disturbance Rejection Control
CL	Closed Loop
CT	Continuous Time
DC	Direct Current
DT	Discrete Time
FB	Feedback
FF	Feedforward
ESO	Extended State Observer
ESR	Equivalent Series Resistance
LADRC	Linear Active Disturbance Rejection Control
LHP	Left Half Plane
MAD	Mean Absolute Deviation
MCU	Microcontroller Unit
MPC	Model Predictive Control
NADRC	Nonlinear Active Disturbance Rejection Control
NESO	Nonlinear Extended State Observer
NLSEF	Nonlinear State Error Feedback
NR	Non-realizable
PD	Proportional Derivative
PLC	Programmable Logic Controller
PI	Proportional Integral
PID	Proportional Integral Derivative
PF	Prefilter
PWM	Pulse Width Modulation
RHP	Right Half Plane
RMS	Root Mean Square
SISO	Single Input, Single Output
TD	Tracking Differentiator
TF	Transfer Function
ZOH	Zero-Order Hold
2DOF	Two Degrees of Freedom

Part I
Foundations

Chapter 1
Prelude: A Fresh Look

Abstract Several aspects make this book different, and doing things not the usual way is its recurring motif. Therefore, in this chapter, we focus on establishing those unique aspects of the book that constitute its claimed titular "fresh look." This expression has a double meaning here. On the one hand, it refers to the active disturbance rejection control (ADRC) as this unorthodox approach in the control landscape and on the other to the fresh look at how ADRC can be introduced to those interested in it. We will also answer here some basic important questions like what exactly is in the book, why is it constructed the way it is, to whom the book is for, and what is the best way to use it depending on the reader's background.

What Is This Book About?

This book is on controls. But, as you will soon find out, this is not yet another typical control textbook, if there is even such a thing. Although the basics of control systems were developed in the first half of the twentieth century, time and technology continue to generate refinements. Here, we look at the area of controls from a particular perspective of *active disturbance rejection control* (ADRC). The idea of ADRC, manifested through its concepts, operating principles, and tools, allows its users to have an alternative view of how to look at, analyze, and solve control problems. Since its inception, ADRC gradually matured into a viable industrial control technology and—as evident through recent adoptions by major companies— found its way into numerous application domains. The unique, disturbance-centric perspective of ADRC helps develop high-performance controllers but with their overall complexity comparable to standard PID. The same cannot be said about the majority of so-called advanced control algorithms which were never successfully transitioned from academia to industry. The relative simplicity, high adaptability, robustness, and scalability are reasons ADRC is often mentioned as a potential contender to PID, which has dominated the field for over 100 years.

How Is This Book Organized?

This book is divided into two main parts, following the book's subtitle. The first one is what we consider the theoretical foundations of ADRC. This part intends to give the reader basic information on the subject in a tutorial style. It is meant to be a self-contained, one-stop shop, where the readers can get a solid footing on ADRC without needing to refer to external sources. The second part deals with all the aspects needed to put the ideas, methods, and tools introduced earlier into practice. In the book, one can find the Prelude, Interlude, and Postlude sections serving as framing devices, which, as the names suggest, discuss the opening, midpoint, and closing matters of the book, respectively.

The first, theoretical, foundation-laying part of the book is organized as follows.

- Chapter 2 is the titular *first contact* with ADRC. It is an example-driven, step-by-step, elementary introduction aiming at developing an initial feel for ADRC. Through this chapter, we want to encourage the readers' use of common sense and engineering intuition and thus liberate them from the limiting dichotomies of control theory versus practice and classical versus modern control algorithms. We want to show that alternatives exist, like ADRC, where synergistic solutions can be conceived by taking what is useful from different worlds.
- Chapter 3 generalizes the concepts introduced earlier but does it in a more stripped-down and concise way using the standard control domain-related terminology and a bit more mathematical rigor. The result of the chapter is the commencement of a base form of ADRC, a kernel one can start with when building customized ADRC-based solutions.
- Chapter 4 takes the previously established base ADRC and contextualizes it to deliver interpretations of it from the established field of linear control systems. This allows the positioning of ADRC among the established control methods.
- Chapter 5 showcases ADRC in a predominantly visual manner. Here, we go deep into the capabilities and limitations of ADRC and foster the reader's intuition and understanding of what is actually possible with this control framework.
- Chapter 6 makes a case that the ADRC should be understood as a modular, flexible framework that could be tailored to specific applications. In this chapter, we show some potential extensions and modifications to the ADRC scheme that can help the designer develop customized solutions addressing specific control problems and/or characteristics of the controlled plant.
- Chapter 7 is the concluding chapter of the first, theoretical part of the book. It is thus a good time to take a look back and around. Since ADRC was not conceived overnight, we give credit where credit is due and briefly describe the history of ADRC including explaining how the elements in the previous chapters came to life. This is also the place where we justify why we decided in this book to mainly focus on a particular form of linear ADRC. To widen the readers' horizons in terms of ADRC, in the second half of the chapter, we give a list of proposed further readings addressing various aspects of ADRC not (at least directly) covered in this book.

1 Prelude: A Fresh Look

The second, practical part of the book consists of the following chapters.

- Chapter 8 opens the second part of book. Since putting ADRC in practice will almost always be in the form of a software-based implementation, we present several discrete-time variants of ADRC and discuss their specific features. We present different styles of implementation and show how it can be optimized for a low computational footprint.
- Chapter 9 examines some of the issues one can encounter when implementing a controller in the real world. Here we focus on those aspects that are most commonly seen in practice. At the same time, we offer specific solutions to those problems that retain the simplicity of ADRC, which is one of its trademarks.
- Chapter 10 shows that going from principles to practice is immediately possible based on the results from previous chapters. Here we demonstrate the implementation of discrete-time variants in software form using both a model-based environment and low-level C language code. We take full advantage of the "cooking recipes," which are handy, ready-to-use procedures, introduced throughout the book, that streamline getting the ADRC tuning parameters for all variants in the book.
- Chapter 11 puts ADRC to work after establishing all the necessary components for its practical implementation. Here we utilize laboratory testbeds to mimic real control problems from process and power control areas and apply knowledge from this book to solve them. To make the experiments reproducible by the readers, we limit ourselves only to affordable and off-the-shelf hardware platforms. With this choice, we hope the readers will be able to first go through our hardware examples and then be ready to take ADRC and deploy it in their applications.
- Chapter 12 deals with the future, both nearest, and that a bit further. Regarding the former, we show what is still coming in the book, and we guide the reader to the Appendices. Regarding the latter, we show what the reader can do after reading this book, discuss possible avenues of further development, and speculate which direction ADRC can evolve in.

The substantive part of the book does not end with the numbered chapters. Appendix A is a reference guide for returning readers that repeats and summarizes all relevant equations required to implement and tune all variants of ADRC that have been introduced in this book. Appendix B provides an overview of some of the existing ADRC toolboxes for MATLAB/Simulink, including the "Linear ADRC Blockset" in Sect. B.3, which is our original toolbox facilitating the implementation of ADRC, in particular those variants and formulas covered in this book.

With the above chapter list and descriptions, one can start to see a pattern here. This book is written in a way that the further you go into the book, the more practical and concise it gets. As the book progresses, the previously minutely explained and analyzed theoretical concepts are gradually transformed and boiled down into succinct practical tools and solutions. All this reaches its culmination point in the Appendix, where different variants of ADRC, covered throughout the book, are blatantly distilled into plug-and-play formulae, which one can simply take and directly deploy in embedded systems, on programmable logic controllers (PLCs), etc.

To Whom Is This Book Addressed and How to Use It?

These two questions need to be answered jointly because how to use this book depends on who you are on the control spectrum. First of all, the book is for those who design and tune control loops regularly and those still learning that. It can play different roles while addressing a wide range of users with different levels of knowledge and skills, wishing to discover and learn ADRC. It also aspires to bridge the theory-practice gap and put ADRC into the hands of every control engineer.

- **For students**, it can be a textbook being part of an introductory course in control systems at both senior undergraduate and first-year graduate levels in several engineering departments. Its two-part structure organically creates a two-semester course. The first semester (Chaps. 2–7) could be focused on building the foundations of ADRC from scratch, while the second semester (Chaps. 8–12) could take those fundamental ideas and tools and show how to apply them to practical scenarios. If the book is used as didactic material, it guides the reader through the subject bit by bit, covers a wide range of topics within ADRC, and provides in-depth information and explanations. The same hardware platforms from Chap. 11 could be used in class to aid students in their ADRC learning experience as those are easily accessible and reasonably priced laboratory testbeds.
- **For practitioners**, it can be viewed as a compact guide on how to use ADRC, in other words, an introduction if you already have a control background in methods like PID. For more experienced users, it can be perceived as a handbook of straightforwardly deployable control solutions that offer improved performance over conventional control applications. We wrote the book having a strong belief that theory must be accessible to engineers and that there is little point in offering a theory that, because of its complexity, is unlikely to find its way into engineering practice. While avoiding overly complicated scientific jargon, the book also explains some ready-to-use modifications and extensions that allow practicing engineers to customize their ADRC algorithms and implement solutions that are tailored to the actual real-world control scenarios.
- **For scientists**, the book offers a systematic and carefully distilled body of work, being a solid foundation of ADRC one can build on. Since the theory-practice gap is heavily tackled in the book and even manifested through its two-part structure, scientists can get a good grasp on the current state of things and be able to further explore basic and applied research questions in the area of ADRC.

The book can be used for self-study and as a one-volume reference for both control novices and the already initiated. Having in mind its introductory nature, the prerequisites for studying the considered subject are intentionally not very strict. Regardless of the intended audience, to get a meaningful grasp of the described underlying concepts, methods, and tools in ADRC, we recommend having at least basic knowledge of systems and controls, as well as an understanding of core mathematical concepts of calculus, linear algebra, and differential equations. Basic programming skills, even though not mandatory, could turn out to be helpful, especially in the more practically oriented second part of the book.

What Are This Book's Guiding Principles?

With this book, our goal is to offer a "fresh look" at how ADRC can be introduced. So if at any point in reading the book, you get a sense that it is peculiar in terms of our choices of language, structure, and content, then most likely it is because we tried to align all elements of the book to the below main motives. To us, this is the book's strength and one of its selling points.

- **Self-containment**: You may have noticed already that we are almost at the end of Chap. 1 and we have not yet provided a single literature reference, which is not typical for the first chapter of a book. In fact, you will not encounter any references until the end of Part I. Such a different use of literature is not a coincidence but rather this book's leitmotiv. It is done to not distract you from the story we are telling and to give you an uninterrupted introduction to ADRC. It is on purpose to make the book as self-contained as possible. We deliberately delay giving credit, context, and historical perspective to Chap. 7, which serves as a thematic midpoint where all the literature references are concentrated and briefly discussed. But even then, there is no need to read any of the papers or books mentioned there to get from start to finish as the book aims to stand on its own. The discussed references should be thus treated as useful pointers rather than prerequisites. Our goal for writing the book was to provide a relatively complete look around the ADRC area without external dependencies besides fundamentals of control systems (like state space and transfer functions).
- **Multiplicity of entry points**: We have structured and written the book to facilitate not only who you are in the control field but also what is your prior experience with ADRC. This means that how to find your way through the book depends on your initial knowledge of ADRC. We offer several entry points where a person can get acquainted with the core concepts of the book. We try to pick people up where they are in terms of experience and knowledge and through a customized guide give here some hints and personal recommendations of what may be of particular interest. Given the fact that the book is intended for multiple audiences, it is important for us to help the readers find their way through the book and quickly filter the information they truly seek.
- **Practicality and reproducibility**: Everything that went into this book had to positively answer the question: Is it practically useful? Such a selection process of what to include was applied to the presented ADRC variants, tuning methodology, methods of implementation, etc. This was our litmus test for deciding whether something contributes to bridging the theory-practice gap in control, which we aim for. That is why we deliberately keep scientific jargon in the book to a minimum and focus on implementability as we want the reader to try ADRC out for oneself. That is also why we provide ready-to-use "cooking recipes" for different ADRC variants in the book. One can easily take those and try them out in application. Furthermore, in the experimental section of the book, we use cheap and commercially available target platforms, so anyone can reproduce all of the results seen in this book, including the hardware ones.

- **Lasting value**: This book is not intended to be read only once. We expect it to be revisited, whenever deemed necessary. When reading it for the first time, one probably moves intuitively from front to back. But when one returns to the book, it can also be read starting at the end as a way to recap the key practical takeaways. For those who have already gone through the book, you may want to go to the Appendix upon subsequent reads, where we have condensed and summarized the material. You can then use the book as a reference manual, designed to be consulted for information on a specific topic and organized in a way to help locate specific information quickly. That is why we go the extra mile in the book by offering toward the end some specially arranged information like the ADRC software implementation or the MATLAB/Simulink blockset that has the ADRC variants covered in the book ready to be deployed.

In Fig. 1.1, we provide some subjective suggestions to readers on what to read next. For the newcomers to the topic of ADRC, we recommend starting with Chap. 2, where we begin explaining ADRC without any assumptions (besides basic control fundamentals) and then proceeding with the book in chronological order. For those who already have some knowledge of ADRC and know the terminology usually associated with it, we suggest starting to find out what this book is about in Chap. 3.

For returning readers, we provide several potential paths, the choice of which depends on what particular aspect of ADRC the reader wants to revisit.

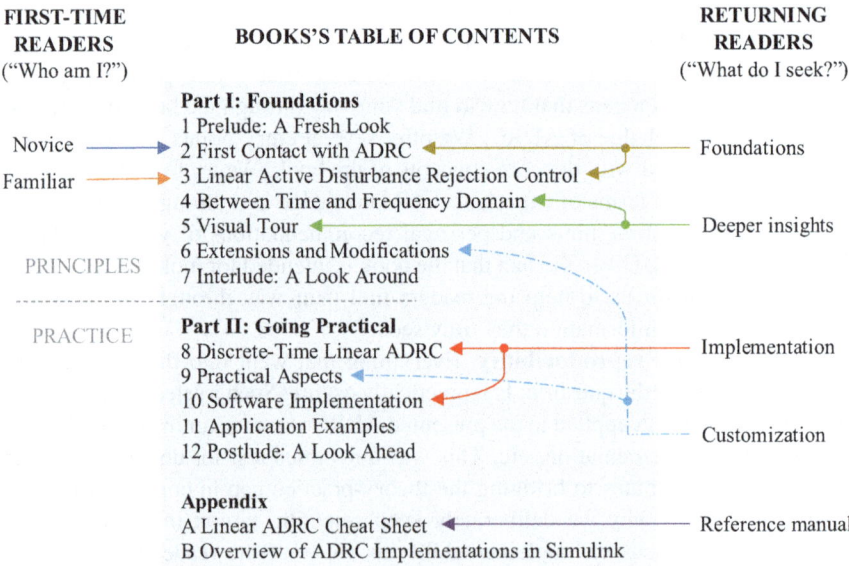

Fig. 1.1 Reading guidance with proposed entry points for the book's first-time readers (left-hand side) and for those who return to the book seeking a specific topic (right-hand side)

Open Access This chapter is licensed under the terms of the Creative Commons Attribution 4.0 International License (http://creativecommons.org/licenses/by/4.0/), which permits use, sharing, adaptation, distribution and reproduction in any medium or format, as long as you give appropriate credit to the original author(s) and the source, provide a link to the Creative Commons license and indicate if changes were made.

The images or other third party material in this chapter are included in the chapter's Creative Commons license, unless indicated otherwise in a credit line to the material. If material is not included in the chapter's Creative Commons license and your intended use is not permitted by statutory regulation or exceeds the permitted use, you will need to obtain permission directly from the copyright holder.

Chapter 2
First Contact with ADRC

Abstract We strive to make the initial encounter with ADRC a smooth and pleasant experience for novices. Therefore, in this chapter, we will rely heavily on illustrative examples inspired by real plants. With such a "show, don't tell" narration, we want to introduce the fundamental aspects of ADRC for some easily recognizable systems from engineering practice. We purposefully avoid complicated jargon and derivations here to give readers the gist of how the ADRC approach can be used to look at and solve various control problems. We deliberately focus on showing *ad hoc* solutions for first- and second-order plants as they are the most prevalent. Through the shown illuminative examples, we hope that you will not only start seeing a pattern in the design process of the ADRC methodology but also be intrigued enough about its simplicity and effectiveness to want to explore it further with this book.

2.1 Control of First-Order Plants

2.1.1 It All Starts with a Model—Or Does It?

For many control methods, especially the more modern and powerful ones, a (more or less) detailed model of the plant is necessary. Even the ubiquitous members of the PID controller family, who continue to persist in industrial practice despite decades of academic efforts, require some knowledge about the process. The character of such knowledge may vary as vastly as its sources, however, and range from theoretical considerations, leading to differential equations, to empirical or experimental data, gathered in time or frequency domain, or, maybe at the very low end, at least a rough idea of where to start experimenting in the search space of controller parameters.

Our goal for this section is to learn about the role of a plant model when using ADRC and how it is positioned in this landscape outlined so far. To make this more tangible, we will start by considering concrete examples.

Level Control of a Water Tank

Imagine your task is to design a controller for the plant sketched in Fig. 2.1, e.g., as part of a production process. The control objective is to keep the level (h) of the liquid in the tank at a desired level (h_r) by adjusting the inlet valve position (v).

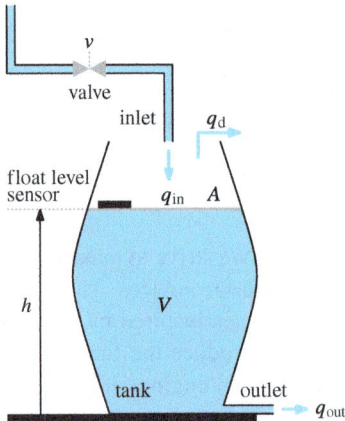

Fig. 2.1 Example of a hydraulic plant, where h is the liquid level (measurable with a float level sensor), v the inlet valve position, q_{in} the inlet flow, q_{out} the outlet flow, A the (varying) cross-section of the tank, and V the current volume of liquid in the tank. The term q_d may represent level changes due to accidental spillage, evaporation, etc.

As the underlying physics is relatively simple to understand, we can develop a mathematical model of this process from first principles. Using a volume balance equation known from fluid mechanics, one can write

$$\dot{V}(t) = q_{in}(t) - q_{out}(t) - q_d(t). \tag{2.1}$$

With a simplifying assumption that the inlet flow is proportional to its valve position, we can substitute $q_{in} = c \cdot v$, with c being the gain value. If the outlet flow depends only on hydrostatic pressure, we can intuitively express it as a function of level h and some parameters $\boldsymbol{p}$, yielding $q_{out} = Q(h, \boldsymbol{p})$. And thirdly, volume V is related to h by the level-dependent cross-section area, i.e., $\dot{V}(t) = A(h) \cdot \dot{h}(t)$. Putting this together allows us to rewrite (2.1) alternatively as

$$A(h) \cdot \dot{h}(t) = c \cdot v(t) - Q(h(t), \boldsymbol{p}) - q_d(t). \tag{2.2}$$

In control jargon, h is the controlled variable (often denoted as y), h_r the set point or reference value (denoted as r), and v the manipulated variable (denoted as u), and $q_d(t)$ acts as an input disturbance (denoted as d) to the system. Rearranging (2.2) and replacing the variable names according to their role in a control loop, we obtain

$$\dot{y}(t) = -\frac{Q(y(t), \boldsymbol{p})}{A(y(t))} - \frac{1}{A(y(t))} \cdot d(t) + \frac{c}{A(y(t))} \cdot u(t). \tag{2.3}$$

To make the derived theoretical model (2.3) useful for controller design, we need to know not only its structure but also its ingredients. This is easier said than done,

2.1 Control of First-Order Plants

and there are several sources of modeling uncertainties. For example, we linearly approximated the inlet flow behavior. Even more importantly, $A(y(t))$ and $Q(y(t), \boldsymbol{p})$ are both unknown functions that change nonlinearly with the operating point. The derived analytical description (2.3) is therefore not complete. Enhancing this model to be close to reality will require much more time and expert knowledge.

Let us therefore have an alternative take at rearranging (2.2), now grouping all unknown elements as a new, "virtual" disturbance input $f(t)$. This includes the nonlinear terms Q and A, the actual disturbance q_d, and an inevitable modeling error regarding the input gain, which we denote as $\Delta b \cdot v(t)$. You might be worried that this convenience comes at a price, but we will soon provide a solution that allows to get rid of these unknowns altogether. Our final plant model now reads

$$\underbrace{\dot{h}(t)}_{\dot{y}(t)} = \underbrace{-\frac{Q(h(t), \boldsymbol{p})}{A(h(t))} + \Delta b \cdot v(t) - \frac{1}{A(h(t))} \cdot q_\mathrm{d}(t)}_{f(t)} + \underbrace{\frac{c}{A(h(t))}}_{b_0} \cdot \underbrace{v(t)}_{u(t)}. \quad (2.4)$$

Structurally, we obtained a very simple disturbed first-order model of our plant: $\dot{y}(t) = f(t) + b_0 \cdot u(t)$. Before we continue to learn how this helps to create a control system, let us have a look at a second concrete example.

Temperature Control of a Heater

Temperature control is a task often not only found in process industries but also encountered in everybody's daily life. To become acquainted with the underlying physics, we will first consider control of the temperature ϑ of an object with mass m in a generic fashion. A heating source shall apply thermal power Q to this object. Assume that heat transfer mechanisms with the environment can fully be described using convection and radiation. The Stefan-Boltzmann law states that the total power radiated from an object with a surface area A is $A\varepsilon\sigma \cdot (\vartheta^4 - \vartheta_\mathrm{a}^4)$, where ε is emissivity, σ is the Stefan-Boltzmann constant, and ϑ_a is the ambient temperature. The convective heat transfer rate is $AU \cdot (\vartheta - \vartheta_\mathrm{a})$, where U is the overall heat transfer coefficient. Putting this together for our generic heated object, having a specific heat capacity c_p, we obtain for the rate of change of the object's thermal energy:

$$mc_\mathrm{p} \cdot \dot{\vartheta}(t) = AU \cdot (\vartheta_\mathrm{a} - \vartheta(t)) + A\varepsilon\sigma \cdot \left(\vartheta_\mathrm{a}^4 - \vartheta(t)^4\right) + Q(t). \quad (2.5)$$

Using control system language and symbols, ϑ acts as the controlled variable (y), and Q is the manipulated variable (u). The ambient temperature is a disturbance to the system, which could, with additional effort and costs, be measured using a second sensor. Accurately obtaining the physical parameters could turn out to be very difficult. Given that (2.5) is only a lumped-parameter model of a real-world plant, its accuracy will, however, always be limited. Its overall structure is similar to (2.2). Both are processes that can be modeled with first-order differential equations. An imbalance in inflows and outflows defines the rate of change in their quantities.

Similar to the water-tank system, let us, therefore, rearrange (2.5) such that unknowns and uncertainties, such as errors in the input gain (Δb), are combined in a virtual disturbance $f(t)$. This will also contain the ambient temperature, so that we do not have to measure it in addition to our actual controlled variable $y(t) = \vartheta(t)$. We then obtain for the plant model

$$\underbrace{\dot{\vartheta}(t)}_{\dot{y}(t)} = \underbrace{\frac{AU}{mc_p} \cdot (\vartheta_a - \vartheta(t)) + \frac{A\varepsilon\sigma}{mc_p} \cdot (\vartheta_a^4 - \vartheta(t)^4) + \Delta b \cdot Q(t)}_{f(t)} + \underbrace{\frac{1}{mc_p}}_{b_0} \cdot \underbrace{Q(t)}_{u(t)}.$$

Being in the form $\dot{y}(t) = f(t) + b_0 \cdot u(t)$, the structure of this model is now very similar to (2.4). Let us get more practical and consider the example of heating a room. Even if we declared most of the model details to be part of a disturbance $f(t)$, we still have to obtain the physical parameters needed for our remaining input gain b_0. An alternative solution is resorting to a simple experiment: the well-known step response.

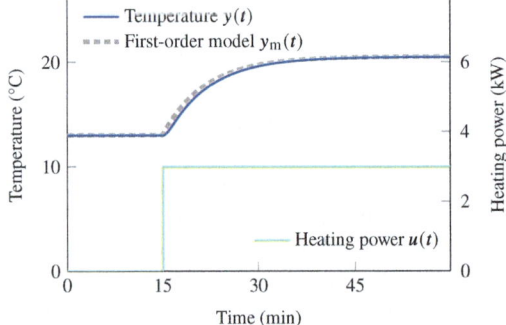

Fig. 2.2 Room temperature step response to a change in heating power (simulated). The dynamics can very well be captured by a linear first-order model, as indicated

Figure 2.2 shows an example of a possible response. As can be seen, a simple first-order model captures the relevant dynamics of this plant very well. In this case, its transfer function is

$$P(s) = \frac{y(s)}{u(s)} = \frac{K}{Ts + 1} \approx \frac{2.5 \, \frac{°C}{kW}}{7 \min \cdot s + 1}. \qquad (2.6)$$

But how can we obtain the value of b_0? We find out by rearranging the differential equation for a generic first-order system to a now already familiar form:

$$T \cdot \dot{y}(t) + y(t) = K \cdot u(t) \quad \rightarrow \quad \dot{y}(t) = \underbrace{-\frac{1}{T} \cdot y(t)}_{f(t)} + \underbrace{\frac{K}{T}}_{b_0} \cdot u(t).$$

2.1 Control of First-Order Plants

This result reveals that we can apply our approach of simplifying the plant model to $\dot{y}(t) = f(t) + b_0 \cdot u(t)$ to other first-order plants, with b_0 often being trivial to obtain, e.g., from gain and time constant parameters.

Bottom Line

Taking a detour from our initial example brought us the following insight: Not only did we find a suitable simplified model for a whole class of problems, namely processes with a balance equation $\frac{d}{dt}$(quantity) = ($\sum$ intake − $\sum$ expenditure) at their heart, but also for any process that can reasonably well be described as a first-order plant $P(s) = \frac{K}{Ts+1}$. Our model has the structure of an integrator with an input disturbance:

$$\dot{y}(t) = f(t) + b_0 \cdot u(t). \tag{2.7}$$

The only remaining parameter is b_0, and we can obtain it from either theoretical analyses or experimental data, building on established methods of system identification—the choice of which depends on access to expert knowledge, capability of conducting experiments, and availability of resources. If we now create a control solution for this simplified model, we will be able to control all of these plants with the same small modeling effort. Time to move on!

2.1.2 Getting Rid of the Disturbance

In Sect. 2.1.1 we substantially simplified the modeling of first-order plants by aggregating elements that are mostly unknown or characterized by high levels of uncertainty into a new, virtual disturbance input $f(t)$ to a first-order integrator (we will later say order $N = 1$). This leaves only one parameter b_0 to be determined.

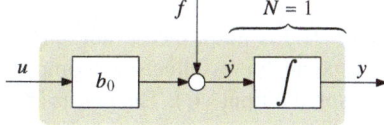

Fig. 2.3 Block diagram interpretation of plant model (2.7) with its interfaces

As might be more apparent from the graphical representation in Fig. 2.3, simplifying the model comes at the price of introducing a second input to an integrator. This appears to complicate things, as it is a bit harder to control $y(t)$ using $u(t)$ if the effects of an unknown and unmeasurable input $f(t)$ must be compensated. Yet this can be achieved in multiple ways, as, after all, the rejection of disturbances belongs to the core competencies of automatic control.

A technique that allows us to separate the tasks of rejecting $f(t)$ and designing the actual controller for $y(t)$ is to let an observer create an estimate $\hat{f}(t)$ of $f(t)$ that allows us to cancel out its effects. With a thoughtful choice of $u(t)$, this will enable a drastic simplification of our actual control task for $y(t)$:

$$\underbrace{\dot{y}(t) = f(t) + b_0 \cdot u(t)}_{\text{disturbed integrator (2.7)}} \overset{u?}{\rightsquigarrow} \underbrace{\dot{y}(t) = u_0(t)}_{\text{pure integrator}}. \tag{2.8}$$

Controlling an undisturbed, pure integrator with a dedicated control action $u_0(t)$ is an easy task, which we will solve in Sect. 2.1.3. It becomes evident from (2.8) that, to achieve the desired transformation, the following overall control action has to be taken, which combines disturbance rejection and control of the integrator:

$$u(t) = \frac{u_0(t) - \hat{f}(t)}{b_0}. \tag{2.9}$$

Constructing this disturbance rejection loop is the crucial step of the entire ADRC approach. Making modeling *and* controller design easier is only possible if rejecting $f(t)$ is done in real time—or, more accurately, faster than the outer, actual control loop. But do not be scared, we can rely on the well-established tool of a linear Luenberger observer to create a sufficiently accurate estimation $\hat{f}(t)$. To do so, we need to create a state-space representation of our model. Even though (2.7) is a first-order system, we make a conscious decision to extend our system description in state space by a second state variable. We select the following state variables:

$$x_1(t) \triangleq y(t), \quad x_2(t) \triangleq f(t).$$

Using these state variables, (2.7) can now be written in state-space form as

$$\begin{pmatrix} \dot{x}_1(t) \\ \dot{x}_2(t) \end{pmatrix} = \begin{pmatrix} 0 & 1 \\ 0 & 0 \end{pmatrix} \cdot \begin{pmatrix} x_1(t) \\ x_2(t) \end{pmatrix} + \begin{pmatrix} b_0 \\ 0 \end{pmatrix} \cdot u(t) + \begin{pmatrix} 0 \\ 1 \end{pmatrix} \cdot \dot{f}(t),$$
$$y(t) = \begin{pmatrix} 1 & 0 \end{pmatrix} \cdot \begin{pmatrix} x_1(t) \\ x_2(t) \end{pmatrix}. \tag{2.10}$$

Since $\dot{f}(t)$ is not a measurable input, the observer for the system (2.10) will have to account for its effects by comparing the estimated system output $\hat{y}(t)$ with the measured output $y(t)$. The observer is governed by the following equations:

$$\begin{pmatrix} \dot{\hat{x}}_1(t) \\ \dot{\hat{x}}_2(t) \end{pmatrix} = \begin{pmatrix} 0 & 1 \\ 0 & 0 \end{pmatrix} \cdot \begin{pmatrix} \hat{x}_1(t) \\ \hat{x}_2(t) \end{pmatrix} + \begin{pmatrix} b_0 \\ 0 \end{pmatrix} \cdot u(t) + \underbrace{\begin{pmatrix} l_1 \\ l_2 \end{pmatrix} \cdot (y(t) - \hat{y}(t))}_{\text{correction term}},$$
$$\hat{y}(t) = \begin{pmatrix} 1 & 0 \end{pmatrix} \cdot \begin{pmatrix} \hat{x}_1(t) \\ \hat{x}_2(t) \end{pmatrix}. \tag{2.11}$$

In (2.11), l_1 and l_2 are observer gains that must be designed together with the outer control loop. For now, let us only sketch the general idea, which is based on pole placement. Combining the equations in (2.11), we obtain

2.1 Control of First-Order Plants

$$\begin{pmatrix} \dot{\hat{x}}_1(t) \\ \dot{\hat{x}}_2(t) \end{pmatrix} = \begin{pmatrix} -l_1 & 1 \\ -l_2 & 0 \end{pmatrix} \cdot \begin{pmatrix} \hat{x}_1(t) \\ \hat{x}_2(t) \end{pmatrix} + \begin{pmatrix} b_0 \\ 0 \end{pmatrix} \cdot u(t) + \begin{pmatrix} l_1 \\ l_2 \end{pmatrix} \cdot y(t),$$

from which we immediately obtain the system matrix of the closed-loop observer. Its pole locations, i.e., eigenvalues λ_1 and λ_2, will determine the observer dynamics—a pragmatic choice will be introduced in Chap. 3. From the characteristic polynomial, we can easily read off equations for l_1 and l_2 in order to achieve the desired dynamics:

$$\det\left(\lambda \mathbf{I} - \begin{pmatrix} -l_1 & 1 \\ -l_2 & 0 \end{pmatrix}\right) = \lambda^2 + l_1 \cdot \lambda + l_2 \stackrel{!}{=} (\lambda - \lambda_1) \cdot (\lambda - \lambda_2) = \lambda^2 \underbrace{-(\lambda_1 + \lambda_2)}_{l_1} \cdot \lambda + \underbrace{\lambda_1 \lambda_2}_{l_2}.$$

Bottom Line

Using an observer is nothing special, but the plant model (2.10) set up for the observer departs from usual practice: Regardless of actual system characteristics, it is always being modeled as a disturbed integrator! The benefit of this particular model is that with $\hat{x}_2(t) = \hat{f}(t)$ we now have the most important ingredient of (2.9) and are able to reject $f(t)$—i.e., both actual disturbances and modeling errors we made.

2.1.3 Control Loop

Once we have successfully set up the disturbance rejection as described in Sect. 2.1.2, designing the outer control loop for the controlled variable $y(t)$ becomes substantially easier. Assuming accurate estimation $\hat{f}(t) = f(t)$, applying the control action (2.9) to (2.7) will lead to a first-order integrator:

$$\dot{y}(t) = u_0(t). \tag{2.12}$$

Our remaining task is then to design the control input u_0 for the integrator dynamics (2.12). A simple proportional (P) controller with a gain value k_1 suffices to follow a reference signal $r(t)$:

$$u_0(t) = k_1 \cdot (r(t) - y(t)). \tag{2.13}$$

Putting equation (2.13) in (2.12) gives us first-order proportional dynamics for the closed loop:

$$\dot{y}(t) = u_0(t) = k_1 \cdot (r(t) - y(t)) \quad \leadsto \quad \frac{1}{k_1} \cdot \dot{y}(t) + y(t) = r(t). \tag{2.14}$$

The system (2.14) having a pole at $-\frac{1}{k_1}$ means we can directly set k_1 to a closed-loop bandwidth (which is the reciprocal value of the closed-loop time constant) straight from design goals and are done with tuning!

Bottom Line

The added slight complexity in the disturbance estimation and rejection phase paid off big time as the outer control loop design was made significantly easier. The required gain value of a simple P controller can be set directly from design goals such as bandwidth or time constant of the closed-loop dynamics. After designing the outer loop, the observer gains in (2.11) can be computed, e.g., by pole placement such that the observer bandwidth is fast enough compared to the closed loop to deliver estimates of the disturbance in a timely manner. We will provide full details on that later in Chap. 3.

Note that we can—and subsequently will—also make use of the output value $\hat{x}_1(t) = \hat{y}(t)$ estimated by the observer, instead of using measurement $y(t)$ in (2.13). This will add some (usually beneficial) low-pass filtering, reducing measurement noise. Putting this all together, we obtain the final control law for the ADRC approach, which employs values estimated by the observer (2.11):

$$u(t) = \frac{1}{b_0} \cdot (k_1 \cdot (r(t) - \hat{x}_1(t)) - \hat{x}_2(t)). \tag{2.15}$$

2.1.4 A First Design Example

Let us apply what we learned by designing a controller for the room temperature example, as we already have concrete model parameters available from Fig. 2.2.[1]

- From (2.6), we know that our only required plant model parameter is

$$b_0 \approx \frac{2.5}{7} \frac{°C}{kW \cdot min} = \frac{2.5}{420} \frac{°C}{kW \cdot s}.$$

- Assuming that there is some headroom left in the heating power, we try to achieve the desired room temperature faster than in the open-loop mode, reducing the time constant from 7 min to 4 min, and accordingly design the controller:

$$k_1 = \frac{1}{4 \, min} = \frac{1}{240 \, s}.$$

- The observer poles are placed five times farther to the left (we will provide all equations later in Chap. 3—it will be equally easy, promised!).

The results presented in Fig. 2.4 do, on the one hand, confirm that we achieved our design goals—quite impressive regarding the little effort we had to put into this. On the other hand, there is clearly room for improvement. Could the response still be sped up? And what about those visible oscillations? Spoiler: In Chap. 11, we will implement a real-world controller for a temperature control task. It turns out that

---

[1] For readers who would like to try to "get their feet wet" with water-tank control instead, a ready-made simulation environment is described in Appendix B.1.

2.2 Towards Control of Higher-Order Systems

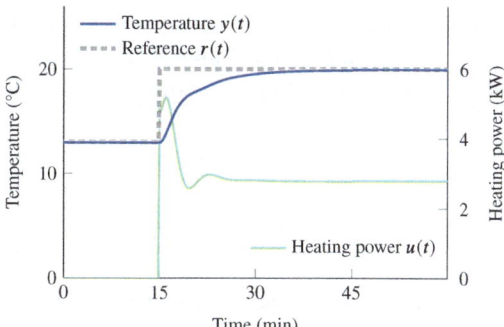

Fig. 2.4 Closed-loop control of the room temperature example from Fig. 2.2

even for such a seemingly simple example, and independent of using ADRC or PI control, there are a lot of details to get right, especially when putting things into practice: Respect the limits of actuators. Take care of higher-order dynamics or dead times to prevent oscillations. Turn on closed-loop control gracefully. And, oh, what about measurement noise?

Well, it is good that not everything is finished yet at the beginning of this book. We are just starting out, with a lot of interesting material to be discovered in the upcoming chapters.

2.2 Towards Control of Higher-Order Systems

The world is not made up only of first-order problems. Let us broaden our horizon and, once more starting from a concrete example, extend our portfolio to second-order systems. Before introducing ADRC for the general case in Chap. 3, this will allow us to develop an intuition on how the approach outlined so far can be gradually extended when being applied to systems with higher-order dynamics.

2.2.1 Plant Modeling of a DC-DC Converter

Departing from the world of rather slow systems, we will consider an example from the power electronics domain here: a synchronous step-down (buck) DC-DC converter. From the simplified schematic given in Fig. 2.5, a plant model shall be

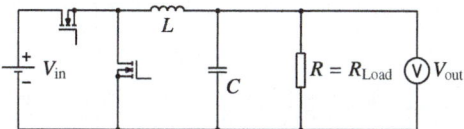

Fig. 2.5 Simplest possible schematic of a synchronous buck DC-DC converter with resistive load. Switches and passive components are assumed to be ideal

derived. Let $i_L(t)$ be the time-dependent inductor current and $v_C(t)$ the capacitor voltage, which equals the output voltage $v_{\text{out}}(t) = v_C(t)$ in this simplified scenario. During the on-time of the high-side switch, the following set of equations holds:

$$\dot{i}_L(t) = \frac{1}{L} \cdot (V_{\text{in}} - v_{\text{out}}(t)) = \frac{1}{L} \cdot V_{\text{in}} - \frac{1}{L} \cdot v_C(t),$$

$$\dot{v}_C(t) = \frac{1}{C} \cdot \left(i_L(t) - \frac{1}{R} \cdot v_{\text{out}}(t) \right) = \frac{1}{C} \cdot i_L(t) - \frac{1}{RC} \cdot v_C(t). \quad (2.16)$$

When the low-side switch is conducting, only the first equation changes:

$$\dot{i}_L(t) = \frac{1}{L} \cdot (-v_{\text{out}}(t)) = -\frac{1}{L} \cdot v_C(t). \quad (2.17)$$

In each cycle of the switching converter, the relative on-time of the high-side switch—also called *duty cycle*—shall be denoted as $\delta(t)$. Using the method of *state-space averaging*, a single set of equations describing the averaged behavior of the converter can be obtained by computing a weighted average of the on- and off-time equations (2.16) and (2.17) (with weights $\delta(t)$ and $1 - \delta(t)$, respectively):

$$\begin{pmatrix} \dot{i}_L(t) \\ \dot{v}_C(t) \end{pmatrix} = \begin{pmatrix} 0 & -\frac{1}{L} \\ \frac{1}{C} & -\frac{1}{RC} \end{pmatrix} \cdot \begin{pmatrix} i_L(t) \\ v_C(t) \end{pmatrix} + \begin{pmatrix} \frac{V_{\text{in}}}{L} \\ 0 \end{pmatrix} \cdot \delta(t). \quad (2.18)$$

With $v_{\text{out}} = v_C$ we can compute a transfer function $P(s)$ connecting duty cycle δ and output voltage v_{out} from the Laplace transform of (2.18) and obtain

$$P(s) = \frac{v_{\text{out}}(s)}{\delta(s)} = (0 \ 1) \cdot \left(s\mathbf{I} - \begin{pmatrix} 0 & -\frac{1}{L} \\ \frac{1}{C} & -\frac{1}{RC} \end{pmatrix} \right)^{-1} \cdot \begin{pmatrix} \frac{V_{\text{in}}}{L} \\ 0 \end{pmatrix} = \frac{V_{\text{in}}}{LCs^2 + \frac{L}{R}s + 1}. \quad (2.19)$$

Clearly (2.19) has the structure of a plant with second-order low-pass behavior. Let us therefore continue in a generalized manner for a second-order plant with input $u(t)$, output $y(t)$, and system parameters K (DC gain), D (damping coefficient), and T (time constant). Its transfer function and differential equation are

$$\frac{y(s)}{u(s)} = \frac{K}{T^2 s^2 + 2DTs + 1} \quad \circ\!\!-\!\!\bullet \quad T^2 \cdot \ddot{y}(t) + 2DT \cdot \dot{y}(t) + y(t) = K \cdot u(t). \quad (2.20)$$

In the spirit of our first-order examples, we rearrange the differential equation for the highest derivative and group the right-hand side as in (2.7), i.e., introducing a "virtual" disturbance $f(t)$ (which may, of course, also contain actual disturbances):

2.2 Towards Control of Higher-Order Systems

$$\ddot{y}(t) = \underbrace{-\frac{2D}{T^2} \cdot \dot{y}(t) - \frac{1}{T^2} \cdot y(t)}_{f(t)} + \underbrace{\frac{K}{T^2}}_{b_0} \cdot u(t). \tag{2.21}$$

For our DC-DC converter example, the value of the model parameter would be $b_0 = \frac{V_{in}}{LC}$. It is a welcome by-product of our derivations that now we already know how to model a widespread class of second-order systems based on system parameters that can be obtained either from analytical modeling or experimentally using established tools of system identification. Comparing (2.21) to (2.7), we can recognize that for the second-order case, only the order of the derivative of $y(t)$ has increased, and the simplified second-order plant model for ADRC is

$$\ddot{y}(t) = f(t) + b_0 \cdot u(t). \tag{2.22}$$

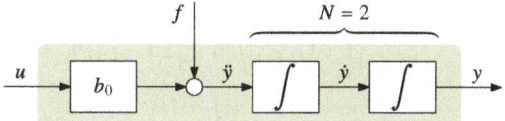

Fig. 2.6 Block diagram interpretation of plant model (2.22) with its interfaces

2.2.2 Inner Loop: Disturbance Rejection and Plant Gain Inversion

Since the simplified plant model (2.22) is so similar to (2.7), which is also visible from its block diagram representation in Fig. 2.6, we can effortlessly apply the same ideas from Sect. 2.1.2—constructing the control action $u(t)$ as a combination of rejecting the disturbance $f(t)$ and compensating the plant gain b_0:

$$u(t) = \frac{u_0(t) - \hat{f}(t)}{b_0}.$$

Putting this in plant model (2.22), i.e., disturbed second-order integrator $\ddot{y}(t) = f(t) + b_0 \cdot u(t)$, will—with a sufficiently accurate estimation $\hat{f}(t)$ of $f(t)$—turn it into an undisturbed, unity-gain second-order integrator. This can then be controlled much more easily by a dedicated control action $u_0(t)$ in the outer control loop:

$$\ddot{y}(t) = u_0(t). \tag{2.23}$$

A linear Luenberger observer can once again provide the necessary estimation $\hat{f}(t)$. Compared to Sect. 2.1.2, we will have to increase the order of the underlying plant model by one and choose the following state variables:

$$x_1(t) \triangleq y(t), \quad x_2(t) \triangleq \dot{y}(t), \quad x_3(t) \triangleq f(t).$$

While our first target is $\hat{f}(t)$, having an estimate of $\hat{y}(t)$ will soon prove very beneficial for controlling (2.23). The system model of our second-order integrator disturbed by $f(t)$, i.e., the state-space representation of (2.22), is

$$\begin{pmatrix} \dot{x}_1(t) \\ \dot{x}_2(t) \\ \dot{x}_3(t) \end{pmatrix} = \begin{pmatrix} 0 & 1 & 0 \\ 0 & 0 & 1 \\ 0 & 0 & 0 \end{pmatrix} \cdot \begin{pmatrix} x_1(t) \\ x_2(t) \\ x_3(t) \end{pmatrix} + \begin{pmatrix} 0 \\ b_0 \\ 0 \end{pmatrix} \cdot u(t) + \begin{pmatrix} 0 \\ 0 \\ 1 \end{pmatrix} \cdot \dot{f}(t),$$

$$y(t) = \begin{pmatrix} 1 & 0 & 0 \end{pmatrix} \cdot \begin{pmatrix} x_1(t) \\ x_2(t) \\ x_3(t) \end{pmatrix}. \tag{2.24}$$

As the disturbance is unknown, it can (for constant or slowly changing disturbances) be reconstructed by the observer from the known input $u(t)$ and output $y(t)$ values. The closed-loop equations of the observer for our system (2.24) are

$$\begin{pmatrix} \dot{\hat{x}}_1(t) \\ \dot{\hat{x}}_2(t) \\ \dot{\hat{x}}_3(t) \end{pmatrix} = \begin{pmatrix} 0 & 1 & 0 \\ 0 & 0 & 1 \\ 0 & 0 & 0 \end{pmatrix} \cdot \begin{pmatrix} \hat{x}_1(t) \\ \hat{x}_2(t) \\ \hat{x}_3(t) \end{pmatrix} + \begin{pmatrix} 0 \\ b_0 \\ 0 \end{pmatrix} \cdot u(t) + \underbrace{\begin{pmatrix} l_1 \\ l_2 \\ l_3 \end{pmatrix} \cdot (y(t) - \hat{y}(t))}_{\text{correction term}},$$

$$\hat{y}(t) = \begin{pmatrix} 1 & 0 & 0 \end{pmatrix} \cdot \begin{pmatrix} \hat{x}_1(t) \\ \hat{x}_2(t) \\ \hat{x}_3(t) \end{pmatrix}. \tag{2.25}$$

Compared to the observer (2.11) for the first-order case, we can recognize that (and how) vectors and matrices in (2.25) grew. The pattern of constructing these for the general case will be revealed in Chap. 3. The tuning procedure for the (now three) observer gains l_1, l_2, and l_3 continues to be based on pole placement. The characteristic polynomial of the closed-loop system matrix of our observer (2.25) is

$$\det\left(\lambda \mathbf{I} - \begin{pmatrix} -l_1 & 1 & 0 \\ -l_2 & 0 & 1 \\ -l_3 & 0 & 0 \end{pmatrix}\right) = \lambda^3 + l_1 \cdot \lambda^2 + l_2 \cdot \lambda + l_3.$$

As in the first-order case, its structure allows to trivially read $l_{1/2/3}$ off a characteristic polynomial constructed using three desired eigenvalues (observer pole locations).

2.2.3 Outer Loop: Control of an Integrator Chain

Once the inner disturbance compensation loop is set up, our controller design task is reduced to finding (and tuning) a suitable control law for the pure, undisturbed second-order integrator (2.23). From classical control engineering, it is known that a proportional-derivative (PD) controller can be employed to stabilize a double integrator system. This means that the P controller approach of the first-order case is now extended by additional feedback of the derivative.

2.2 Towards Control of Higher-Order Systems

Computing the true derivative of a real-time signal $y(t)$ is impossible. In PD and PID controllers, there are workarounds to tackle this task. Luckily (but not accidentally), we already have an estimate of $\dot{y}(t)$ available, provided by the observer. If we, just for the first step, assume this estimate to be ideal, a PD-like control law (not applying the derivative to the reference signal $r(t)$) for $u_0(t)$ would be

$$u_0(t) = k_1 \cdot (r(t) - y(t)) - k_2 \cdot \dot{y}(t). \tag{2.26}$$

Putting (2.26) into (2.23) yields the differential equation for the closed loop:

$$\frac{1}{k_1} \cdot \ddot{y}(t) + \frac{k_2}{k_1} \cdot \dot{y}(t) + y(t) = r(t). \tag{2.27}$$

Comparing (2.27) to the equation of a generic second-order system as in (2.20), we can see that not only did we achieve steady-state accuracy (unity DC gain from reference to system output), but we can arbitrarily shape the dynamic response of the closed control loop using the controller gains k_1 and k_2, as well.

In practice, there will of course be limitations to what can be achieved—firstly, we only have an estimate of $\dot{y}(t)$ available. Secondly, the controller may be faced with dynamics diverging from the double integrator, as there are speed limits to the disturbance rejection required to maintain that behavior, as well. This emphasizes the role of the observer within ADRC: It does the "heavy lifting" for us, simplifying both plant modeling and selecting controller parameters, but we need to take care of properly tuning it.

While we started from PD-like control, after putting disturbance rejection and outer loop control based on the estimated values together in one combined control law, we recognize that, with ADRC, we built an observer-based state-space controller:

$$u(t) = \frac{1}{b_0} \cdot (k_1 \cdot (r(t) - \hat{x}_1(t)) - k_2 \cdot \hat{x}_2(t) - \hat{x}_3(t)). \tag{2.28}$$

Using tools from what is often called "modern" control engineering but for a deliberately "wrong" model, and thus creating a general-purpose controller that will be easy to tune—this dichotomy is what may summarize the philosophy of ADRC best.

Finally coming back to our introductory example for second-order systems: What about the control of a DC-DC converter using ADRC? Let us save another simulation example here. Toward the end of this book, in Sect. 11.2, we will tackle voltage control using ADRC in a real-world example, taking care of various practical aspects discussed in the upcoming chapters, and using our own optimized software implementation that we will have developed by then. We hope to see you there!

2.2.4 A Pattern Emerges: What Constitutes ADRC?

Starting from concrete examples, we have become acquainted with ADRC for first- and second-order systems. Before extending that to the general case in Chap. 3, let us extract the most important concepts and ingredients that constitute ADRC.

Simplified Plant Modeling

British statistician George Box wrote the famous line "All models are wrong, some are useful." His point was that one should focus more on whether something can be applied to everyday life in a useful manner rather than debating endlessly if an answer is correct in all cases. From a control point of view, this calls for a control design that is not overly dependent on the mathematical model of the plant, is robust to modeling uncertainties, and at the same time does not require performing tedious experiments identifying details of the plant model. And this is when ADRC comes to play. Let us compare the models for the first- and second-order cases:

$$\dot{y}(t) = f(t) + b_0 \cdot u(t), \qquad (2.29a)$$
$$\ddot{y}(t) = f(t) + b_0 \cdot u(t). \qquad (2.29b)$$

The plant in ADRC is modeled as an integrator of a certain order, perturbed by an external signal (f): a combination of usually unknown or uncertain terms, which can include actual disturbances and modeling errors. Whether theoretical analysis or an empirical approach led to this simple mathematical input-output representation, this model is the core of ADRC—and the one we build a controller around. Furthermore, apart from the number of integrator stages, it has just one parameter (b_0), which makes modeling—and later, tuning the controller—generally easier.

Normalizing the Plant: Disturbance Rejection and Gain Inversion

An inner loop within the ADRC concept is set up to reject the influence of the disturbance (f). In addition, effects of the plant gain (b_0) are being counteracted by applying its inverted value $1/b_0$ at the output of our control signal $u(t)$:

$$u(t) = \frac{u_0(t) - \hat{f}(t)}{b_0}. \qquad (2.30)$$

Provided that b_0 is known and an estimation $\hat{f} \approx f$ is available, both ideas create an inner loop with (approximately) the dynamics of an undisturbed integrator (chain), with an input $u_0(t)$. This becomes visible when putting (2.30) into (2.29):

$$\dot{y}(t) \approx u_0(t), \qquad (2.31a)$$
$$\ddot{y}(t) \approx u_0(t). \qquad (2.31b)$$

2.2 Towards Control of Higher-Order Systems

Estimating the Disturbance and States

To enable rejection of the disturbance, its value must be estimated in real time (as $\hat{f}$), since it usually cannot be measured. This task is assigned to an observer, which, at the same time, also estimates state variables that are being used in the final building block of ADRC: the outer state-feedback controller. Let us compare the condensed observer equations for the first and second cases:

$$\begin{pmatrix} \dot{\hat{x}}_1(t) \\ \dot{\hat{x}}_2(t) \end{pmatrix} = \begin{pmatrix} 0 & 1 \\ 0 & 0 \end{pmatrix} \cdot \begin{pmatrix} \hat{x}_1(t) \\ \hat{x}_2(t) \end{pmatrix} + \begin{pmatrix} b_0 \\ 0 \end{pmatrix} \cdot u(t) + \begin{pmatrix} l_1 \\ l_2 \end{pmatrix} \cdot \left(y(t) - \begin{pmatrix} 1 & 0 \end{pmatrix} \cdot \begin{pmatrix} \hat{x}_1(t) \\ \hat{x}_2(t) \end{pmatrix} \right), \quad (2.32a)$$

$$\begin{pmatrix} \dot{\hat{x}}_1(t) \\ \dot{\hat{x}}_2(t) \\ \dot{\hat{x}}_3(t) \end{pmatrix} = \begin{pmatrix} 0 & 1 & 0 \\ 0 & 0 & 1 \\ 0 & 0 & 0 \end{pmatrix} \cdot \begin{pmatrix} \hat{x}_1(t) \\ \hat{x}_2(t) \\ \hat{x}_3(t) \end{pmatrix} + \begin{pmatrix} 0 \\ b_0 \\ 0 \end{pmatrix} \cdot u(t) + \begin{pmatrix} l_1 \\ l_2 \\ l_3 \end{pmatrix} \cdot \left(y(t) - \begin{pmatrix} 1 & 0 & 0 \end{pmatrix} \cdot \begin{pmatrix} \hat{x}_1(t) \\ \hat{x}_2(t) \\ \hat{x}_3(t) \end{pmatrix} \right).$$
(2.32b)

One can clearly recognize a pattern emerging in the matrices and vectors. The first estimated state is $\hat{x}_1 = \hat{y}$, the final one is the disturbance estimation $\hat{f}$, and the ones in between are estimated derivatives of y.

Accurate and timely estimation is crucial to the success of the ADRC approach. A well-tuned observer is the cornerstone to some of its main selling points: easier plant modeling and easier controller design.

Controlling an Integrator Chain

Enabled by the inner disturbance rejection and plant gain inversion loop, the outer state-feedback controller can now always be designed for a unity-gain integrator chain—which is easy when the internal states are made available by the observer. State-feedback equations for the first- and second-order cases are

$$u_0(t) = k_1 \cdot (r(t) - \hat{x}_1(t)), \quad (2.33a)$$
$$u_0(t) = k_1 \cdot (r(t) - \hat{x}_1(t)) - k_2 \cdot \hat{x}_2(t). \quad (2.33b)$$

Simplified Controller and Observer Tuning

Apart from b_0, which is part of the control law (2.30) and must be obtained during plant modeling, the following controller and observer parameters must be tuned:

- First-order case: k_1 (controller), and l_1, l_2 (observer)
- Second-order case: k_1, k_2 (controller), and l_1, l_2, l_3 (observer)

One additional controller and one additional observer gain value must be tuned when increasing the plant order. How ADRC's inner loop eases controller tuning becomes obvious when looking at the approximation of the closed loop for the first- and second-order cases (approximate since the inner loop will not be perfect in practice):

$$\frac{1}{k_1} \cdot \dot{y}(t) + y(t) \approx r(t), \tag{2.34a}$$

$$\frac{1}{k_1} \cdot \ddot{y}(t) + \frac{k_2}{k_1} \cdot \dot{y}(t) + y(t) \approx r(t). \tag{2.34b}$$

Clearly (2.34) has first- and second-order low-pass behavior, respectively, which allows to derive equations for the controller gains based on desired dynamics of the closed loop.

Although we introduce a concrete tuning approach only in Chap. 3, one can already feel confident that computing observer gains will also be simple, as their characteristic polynomials are $\lambda^2 + l_1 \cdot \lambda + l_2$ (first-order case) and $\lambda^3 + l_1 \cdot \lambda^2 + l_2 \cdot \lambda + l_3$ (second-order case), allowing to directly read off equations for the gains once a desired polynomial is given through pole placement design.

Bottom Line

Bringing all the pieces of ADRC together, we can visualize it through a block diagram shown in Fig. 2.7. Note that we added an optional feedback path of the plant output, indicating that this measured value could be used by the outer controller instead of an estimated version delivered by the observer (in this book, however, we will not).

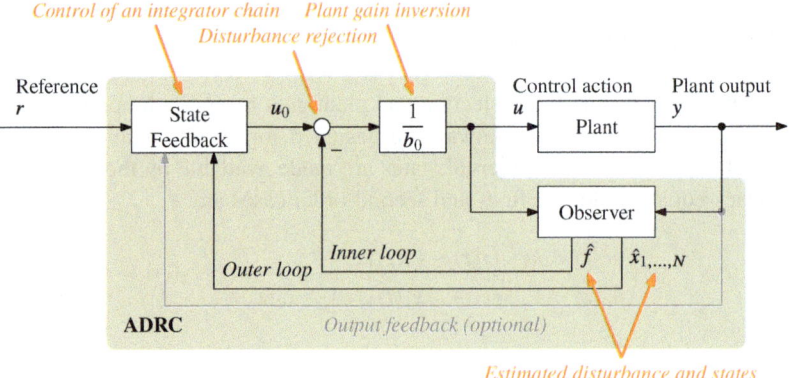

Fig. 2.7 General scheme of a control loop with ADRC and its ingredients

With the benefits mentioned above, there come challenges as well. We do not try to hide complexity. It is clear that a certain workload is needed for designing ADRC, but it is meant to simplify things in the end. In this book, we argue that concerning some of the popular classical and modern control approaches, ADRC is simpler without sacrificing performance. Looking back at how the ADRC has been applied to the first- and second-order plant examples, we started to see a pattern. In the next chapter, we will generalize the observations made here and present the blocks necessary to build ADRC from the ground up.

Open Access This chapter is licensed under the terms of the Creative Commons Attribution 4.0 International License (http://creativecommons.org/licenses/by/4.0/), which permits use, sharing, adaptation, distribution and reproduction in any medium or format, as long as you give appropriate credit to the original author(s) and the source, provide a link to the Creative Commons license and indicate if changes were made.

The images or other third party material in this chapter are included in the chapter's Creative Commons license, unless indicated otherwise in a credit line to the material. If material is not included in the chapter's Creative Commons license and your intended use is not permitted by statutory regulation or exceeds the permitted use, you will need to obtain permission directly from the copyright holder.

Chapter 3
Linear Active Disturbance Rejection Control

Abstract As the core of all other ADRC variants covered in this book, the linear continuous-time state-space form of active disturbance rejection control is introduced in this chapter. Generalizing the first- and second-order cases considered in Chap. 2, the structure of Nth-order linear ADRC is being discussed, consisting of observer and state-feedback controller. Equations for tuning both observer and controller are derived for the so-called *bandwidth parameterization* approach.

3.1 Derivation and Core Concepts

Building on the examples from Chap. 2, we will now present linear active disturbance rejection control for the general Nth-order case. Starting from a (simple) plant model, this will involve setting up an inner control loop to create dynamics that can be easily controlled in a second, outer loop—both of which make use of an observer that will be essential to the ADRC concept.

3.1.1 Modeling the Plant

Consider the single-input, single-output (SISO) plant of order N in the control loop of Fig. 3.1, with output $y(t)$, input $u(t)$, and input disturbance $d(t)$. Assume that one can describe its behavior with the following simple differential equation:

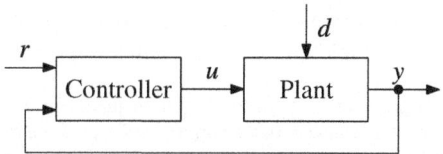

Fig. 3.1 Generic "textbook" structure of a control loop with reference signal r, control signal u (controller output), controlled variable y (plant output), and disturbance d

© The Author(s) 2025
G. Herbst, R. Madonski, *Active Disturbance Rejection Control*, Control Engineering, https://doi.org/10.1007/978-3-031-72687-3_3

$$y^{(N)}(t) + \sum_{i=0}^{N-1} a_i \cdot y^{(i)}(t) = b \cdot u(t) + b_{\mathrm{d}} \cdot d(t). \tag{3.1}$$

When—as mostly the case in practice—the disturbance d is not measurable, the usual remaining assumption for designing a controller using plant model information is that the parameters b and a_i in (3.1) are known. For ADRC, this is not the case; knowledge of the a_i parameters is not needed.[1] A plant model for ADRC only requires the plant order N and an estimate b_0 of the b parameter, typically with an unavoidable error $\Delta b = b - b_0$. Let us rearrange (3.1) for the highest derivative of $y(t)$ such that all unknown terms are conveniently combined to a single time-varying value $f(t)$

$$y^{(N)}(t) = \underbrace{- \sum_{i=0}^{N-1} a_i \cdot y^{(i)}(t) + \Delta b \cdot u(t)}_{\text{modeling errors}} + \underbrace{b_{\mathrm{d}} \cdot d(t)}_{\text{actual disturbance}} + b_0 \cdot u(t),$$

$$\underbrace{\hphantom{- \sum_{i=0}^{N-1} a_i \cdot y^{(i)}(t) + \Delta b \cdot u(t) + b_{\mathrm{d}} \cdot d(t)}}_{\text{total disturbance } f(t)}$$

which, for ADRC, is usually denoted as *generalized* or *total disturbance*, as it consists of unknown model terms (modeling errors) and actual disturbances. Using $f(t)$, we can now see that the plant model used for ADRC has a very compact form:

$$y^{(N)}(t) = f(t) + b_0 \cdot u(t). \tag{3.2}$$

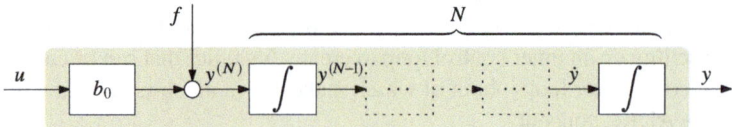

Fig. 3.2 Block diagram interpretation of plant model (3.2) with its interfaces

Being the only required coefficient, a bigger importance is naturally attached to the knowledge of b_0. It is therefore sometimes denoted as the *critical gain parameter* in the context of ADRC. It may also become clear from (3.2) that the starting point of obtaining this model does not have to be a linear equation like (3.1). Other terms and nonlinearities might very well be condensed in $f(t)$, as well. In any case, (3.2) describes the dynamics of an Nth-order integrator chain with input gain b_0 and an additional disturbance input $f(t)$, as shown in Fig. 3.2. Before discovering all the benefits of this, let us conclude our considerations on plant modeling for ADRC with some simple but practically important examples given in Table 3.1.

---

[1] While knowledge of the a_i parameters is not required, it does not have to be thrown away when available. One can indeed improve the performance of the control loop using a more detailed plant model. In Sect. 6.1 we will briefly discuss this topic.

3.1 Derivation and Core Concepts

Table 3.1 Examples of plant model parameters required for ADRC for plants with simple first- and second-order low-pass behavior. If a plant model is available in form of transfer function parameters such as K or T—experimentally obtained by a step response, for example—or in form of a differential equation, this table demonstrates how to obtain required the parameters b_0 (plant gain) and N (plant order)

	First-order low-pass	Second-order low-pass
Transfer function	$\dfrac{y(s)}{u(s)} = \dfrac{K}{Ts+1}$	$\dfrac{y(s)}{u(s)} = \dfrac{K}{T^2 s^2 + 2DTs + 1}$
Differential equation	$\dot{y}(t) = \underbrace{-\dfrac{1}{T}y(t)}_{f(t)} + \dfrac{K}{T}u(t)$	$\ddot{y}(t) = \underbrace{-\dfrac{2D}{T}\dot{y}(t) - \dfrac{1}{T^2}y(t)}_{f(t)} + \dfrac{K}{T^2}u(t)$
Plant model parameters	$b_0 = \dfrac{K}{T}$ and $N = 1$	$b_0 = \dfrac{K}{T^2}$ and $N = 2$

3.1.2 Normalizing the Plant Behavior

It could appear unusual both to treat unknown terms of a plant model as a disturbance and to model a plant as an integrator chain regardless of its actual behavior. If you already had an idea of your plant to be controlled, the model (3.2) might not be what you wanted or expected! So how do these concepts of a *total disturbance* f and *critical gain parameter* b_0 help us in building a controller for our control problem?

Despite having just introduced them, we will now try to get rid of f and cancel out the effects of b_0. This will significantly ease the design of a controller for the closed loop. Firstly, knowledge of b_0 allows scaling the controller output such that the task of tuning the actual control loop becomes independent of plant parameters. Secondly, if an estimate $\hat{f}(t)$ of the *total disturbance* $f(t)$ is available (with a tool we will discuss soon), modeling errors and effects of actual disturbances can be compensated at the same time. Equation (3.2) then gets reduced to a system with well-behaved, known dynamics. The control signal $u(t)$ is therefore set up like this:

$$u(t) = \frac{u_0(t) - \hat{f}(t)}{b_0}. \tag{3.3}$$

Equation (3.3) combines two core ideas of ADRC, whose effects on the plant model (3.2) are visualized in Fig. 3.3:

- *Disturbance rejection:* Feeding $-\frac{1}{b_0} \cdot \hat{f}(t)$ to the plant as part of the control signal $u(t)$ (ideally) compensates the influence of $f(t)$ on the dynamics of (3.2), resulting in an undisturbed chain of N integrators.
- *Plant gain inversion:* Multiplying the control signal component $u_0(t)$ with the reciprocal value of the critical gain parameter b_0 (approximately) creates—together with disturbance compensation—a unity-gain integrator chain with an input $u_0(t)$:

$$y^{(N)}(t) \approx u_0(t). \tag{3.4}$$

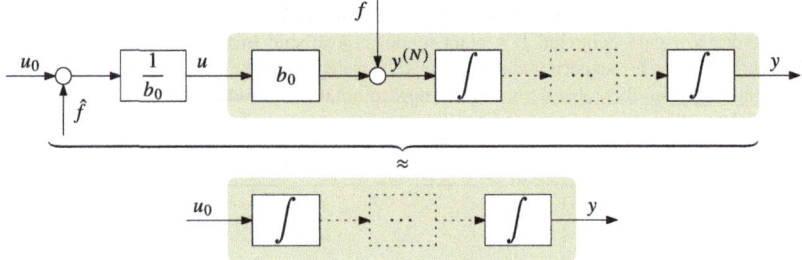

Fig. 3.3 If a sufficiently accurate estimation $\hat{f}$ of f is available and the value of b_0 is known, the control law (3.3) approximately transforms the original plant model (3.2) (disturbed integrator chain) into a unity-gain integrator chain. This can then be controlled much easier using a dedicated control signal component u_0—and controller design will *always* be for an integrator chain, regardless of the actual plant

The required estimate $\hat{f}(t)$ does not appear from nowhere. It will have to be determined using measurements of the real plant, which are then obviously fed back in (3.3). We have therefore just witnessed the creation of a first, inner control loop as part of the ADRC concept. With unity-gain Nth-order integrator chain behavior, this loop behaves like a "normalized plant." Designing a controller with a control signal $u_0(t)$ for that will, as a result, become a straightforward task—we will tackle that right next. Said controller would then be able to work with *any* plant of order N described by (3.2) with the same value b_0 without further modification.

3.1.3 Controlling a Normalized Plant

To better explain the further approach, let us assume the inner loop just introduced in Sect. 3.1.2 works perfectly and now has the behavior of an ideal, undisturbed, unity-gain chain of N integrators:

$$y^{(N)}(t) = u_0(t). \tag{3.5}$$

Let us further assume that the output signals of all integrators, i.e., $y(t)$ and its first $(N-1)$ derivatives, are available. In practice, they are not, but do not worry right now! One can then construct a law for the control signal $u_0(t)$ as follows:

$$u_0(t) = k_1 \cdot (r(t) - y(t)) - k_2 \cdot \dot{y}(t) - \ldots - k_N \cdot y^{(N-1)}(t) \tag{3.6}$$

$$= k_1 \cdot r(t) - \underbrace{\begin{pmatrix} k_1 & \cdots & k_N \end{pmatrix}}_{k^\mathrm{T}} \cdot \begin{pmatrix} y(t) \\ \vdots \\ y^{(N-1)}(t) \end{pmatrix}.$$

3.1 Derivation and Core Concepts

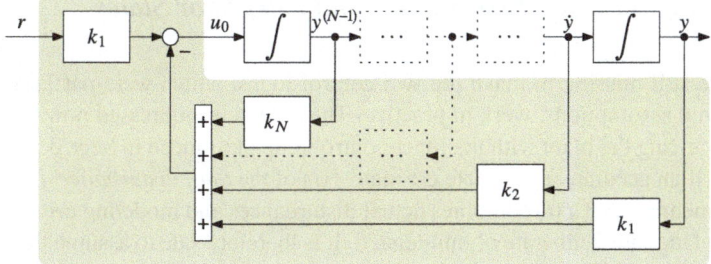

Fig. 3.4 Control of a plant with undisturbed, unity-gain integrator chain behavior based on feeding back the outputs of all integrator stages. The controller is highlighted and also given in (3.6)

Equation (3.4) can be interpreted as a higher-order generalization of a PD controller, with derivatives acting only on the plant output y, but not on the reference signal (set point) r. The resulting loop of control law (3.6) with its plant (3.5) is visualized in Fig. 3.4. In contrast to a traditional PD controller, the derivatives are not computed within the controller, but directly taken from the plant ("measured"). Subsequently, we will therefore adopt the appropriate nomenclature and call (3.6) by its real name: a linear state-feedback controller.

The effectiveness of this approach can be recognized by putting (3.6) into (3.5). With slight rearrangements we obtain the closed-loop behavior as having Nth-order low-pass characteristics and unity DC gain— i.e., a control loop with steady-state accuracy:

$$\frac{1}{k_1} \cdot y^{(N)}(t) + \frac{k_N}{k_1} \cdot y^{(N-1)}(t) + \ldots + \frac{k_2}{k_1} \cdot \dot{y}(t) + y(t) = r(t). \tag{3.7}$$

Remember that the behavior of a "normalized plant" was the result of using an inner control loop as described in Sect. 3.1.2. Within the ADRC concept, we have now built a second, outer loop around that, serving the purpose of shaping the dynamics of the overall control system. What if the inner loop is not perfect, as nothing is? Then (3.5) turns into (3.4), and (3.7) still holds approximately.

Tuning of the gain parameters $k_1, \ldots, k_N$ will be properly addressed in Sect. 3.2.1, but it does not take much imagination that straightforward equations to compute these gains can be obtained from (3.7) if desired dynamics are given in the form of differential equation coefficients for the closed loop. We promised that tuning the outer loop would be easier: The absence of any plant parameters here means that, for ADRC, tuning the controller will be possible independent of such parameters and be directly based only on the desired closed-loop behavior.

This will not free us from the laws of physics and engineering restrictions, of course, which limit the dynamics that can be achieved for a certain plant with a real-world actuator. Nevertheless, this constitutes a departure from the PID world, where controller gains have to be tuned interacting with plant parameters—a very nice quality-of-life improvement for the designer of a control system using ADRC.

3.1.4 Estimating the Total Disturbance and Integrator States

Some pieces are still missing to make the two control loops, which were partially based on idealized assumptions, work in practice. These shall be addressed now.

Firstly, normalizing the plant with the inner control loop introduced in Sect. 3.1.2 is only possible if an accurate, up-to-date estimate $\hat{f}(t)$ of the *total disturbance* $f(t)$ is available. Remember that $f(t)$ combines actual disturbances and modeling errors, regardless if the latter are deliberate or unintended. It is therefore safe to assume that $f(t)$ will never be measurable in practically relevant cases, if only because a plant model will never be capable of capturing all real-world details. All that we usually have available are signals outside of the actual plant: its measured output $y(t)$, and its input $u(t)$, which is the output of our controller, cf. Fig. 3.1. The tool of choice to reconstruct missing signals in control systems engineering is the use of an observer, which continuously provides estimated signal values.

Secondly, after compensating the *total disturbance* within the inner loop, the remaining task is to control a normalized system with integrator chain behavior, using the outer loop described in Sect. 3.1.3. This can—with high flexibility regarding the attainable closed-loop dynamics—very well be solved using a linear state-feedback controller if all outputs of the integrator stages are available. Bringing to mind that the integrator chain is just a (best-case) description of the behavior of the closed inner loop, it becomes obvious that its internal states cannot be measurable. They must be estimated, as well.

These two tasks are being solved jointly using an established tool from the linear control systems world—a Luenberger observer. An observer is built around a model. We already have one that combines an integrator chain with an unknown disturbance: ADRC's plant model (3.2). A state-space description of this is required and can be given as follows:

$$\underbrace{\begin{pmatrix} \dot{x}_1(t) \\ \vdots \\ \dot{x}_{N+1}(t) \end{pmatrix}}_{\dot{x}(t)} = \underbrace{\begin{pmatrix} \mathbf{0}^{N\times 1} & \mathbf{I}^{N\times N} \\ 0 & \mathbf{0}^{1\times N} \end{pmatrix}}_{A} \cdot \underbrace{\begin{pmatrix} x_1(t) \\ \vdots \\ x_{N+1}(t) \end{pmatrix}}_{x(t)} + \underbrace{\begin{pmatrix} \mathbf{0}^{(N-1)\times 1} \\ b_0 \\ 0 \end{pmatrix}}_{b} \cdot u(t) + \begin{pmatrix} \mathbf{0}^{N\times 1} \\ 1 \end{pmatrix} \cdot \dot{f}(t) \qquad (3.8)$$

$$y(t) = \underbrace{\begin{pmatrix} 1 & \mathbf{0}^{1\times N} \end{pmatrix}}_{c^T} \cdot \begin{pmatrix} x_1(t) \\ \vdots \\ x_{N+1}(t) \end{pmatrix}.$$

From c^T in (3.8) it is clear that the first state variable is $x_1(t) = y(t)$. Looking at A furthermore reveals that $\dot{x}_i(t) = x_{i+1}(t)$ holds for $i = 1, \ldots, N-1$, which means that $x_{i+1}(t)$ is the ith derivative of $y(t)$. The next-to-last state equation reads $\dot{x}_N(t) = x_{N+1}(t) + b_0 \cdot u(t)$, corresponding to $y^{(N)}(t) = f(t) + b_0 \cdot u(t)$. This means that $f(t)$ is the final element of the vector of state variables, $x_{N+1}(t) = f(t)$, and consequently, the last line correctly states $\dot{x}_{N+1}(t) = \dot{f}(t)$.

3.1 Derivation and Core Concepts

When setting up a Luenberger observer in (3.9) to obtain an estimate of the states $\hat{\boldsymbol{x}}^T(t) = \begin{pmatrix} \hat{x}_1(t) & \cdots & \hat{x}_{N+1}(t) \end{pmatrix}$, only measurements of $u(t)$ and $y(t)$ are available. Omitting the "virtual input" $\dot{f}(t)$ means $\hat{x}_{N+1}(t)$ is treated as a constant *total disturbance* at the plant model input. That way, $\hat{x}_{N+1}(t)$ serves as the integrator state of ADRC necessary to achieve zero steady-state error for constant reference values and constant disturbances. But of course, just like classical PI or PID controllers, ADRC will also handle time-varying reference signals and disturbances if observer and controller are fast enough. The closed-loop observer equations for ADRC are

$$\underbrace{\begin{pmatrix} \dot{\hat{x}}_1(t) \\ \vdots \\ \dot{\hat{x}}_{N+1}(t) \end{pmatrix}}_{\dot{\hat{\boldsymbol{x}}}(t)} = \underbrace{\begin{pmatrix} \boldsymbol{0}^{N\times 1} & \boldsymbol{I}^{N\times N} \\ 0 & \boldsymbol{0}^{1\times N} \end{pmatrix}}_{A} \cdot \underbrace{\begin{pmatrix} \hat{x}_1(t) \\ \vdots \\ \hat{x}_{N+1}(t) \end{pmatrix}}_{\hat{\boldsymbol{x}}(t)} + \underbrace{\begin{pmatrix} \boldsymbol{0}^{(N-1)\times 1} \\ b_0 \\ 0 \end{pmatrix}}_{b} \cdot u(t) + \underbrace{\begin{pmatrix} l_1 \\ \vdots \\ l_{N+1} \end{pmatrix}}_{l} \cdot (y(t) - \hat{y}(t))$$

$$\hat{y}(t) = \underbrace{\begin{pmatrix} 1 & \boldsymbol{0}^{1\times N} \end{pmatrix}}_{c^T} \cdot \begin{pmatrix} \hat{x}_1(t) \\ \vdots \\ \hat{x}_{N+1}(t) \end{pmatrix}. \quad (3.9)$$

Looking at the dynamics of the estimation error $\boldsymbol{e}_x = \boldsymbol{x} - \hat{\boldsymbol{x}}$ and its derivative, one arrives at the following equation (again ignoring the nonavailable "virtual input"):

$$\dot{\boldsymbol{e}}_x = \frac{d}{dt}(\boldsymbol{x}(t) - \hat{\boldsymbol{x}}(t)) = \dot{\boldsymbol{x}}(t) - \dot{\hat{\boldsymbol{x}}}(t) = \left(A - \boldsymbol{l}\boldsymbol{c}^T\right) \cdot (\boldsymbol{x}(t) - \hat{\boldsymbol{x}}(t)). \quad (3.10)$$

This means that the Luenberger observer gains $\boldsymbol{l}^T = \begin{pmatrix} l_1 & \cdots & l_{N+1} \end{pmatrix}$ must, as part of an ADRC design procedure, be chosen to let the estimation error converge to zero in a stable manner—and quickly enough to provide an estimate of the *total disturbance* for effective disturbance rejection in (3.3), as well as estimates of $y(t)$ and its derivatives required for computing the state-feedback control signal $u_0(t)$. We will cover observer tuning by setting the eigenvalues of $(A - \boldsymbol{l}\boldsymbol{c}^T)$ in Sect. 3.2.2.

If we summarize the signals estimated by the observer,

$$\hat{\boldsymbol{x}}(t) = \begin{pmatrix} \hat{x}_1(t) \\ \hat{x}_2(t) \\ \vdots \\ \hat{x}_N(t) \\ \hat{x}_{N+1}(t) \end{pmatrix} = \begin{pmatrix} \hat{y}(t) \\ \dot{\hat{y}}(t) \\ \vdots \\ \hat{y}^{(N-1)}(t) \\ \hat{f}(t) \end{pmatrix}, \quad (3.11)$$

it is evident that we now have all the signals available to implement the two control loops: $\hat{f}$ for disturbance rejection using (3.3) in the inner loop and estimated derivatives of y for state-feedback control of the integrator chain in the outer loop. For the latter, we can modify the control law (3.6) accordingly:

$$u_0(t) = k_1 \cdot (r(t) - \hat{y}(t)) - k_2 \cdot \dot{\hat{y}}(t) - \ldots - k_N \cdot \hat{y}^{(N-1)}(t). \quad (3.12)$$

3.1.5 Putting It All Together

Combining all previous results from Sect. 3.1, we can now summarize the equations that constitute linear ADRC, using the state variables $\hat{\boldsymbol{x}}(t)$ of the observer listed in (3.11). The inner loop can be realized as

$$u(t) = \frac{u_0(t) - \hat{f}(t)}{b_0} = \frac{u_0(t) - \hat{x}_{N+1}(t)}{b_0}, \qquad (3.13)$$

while the state-feedback control law of the outer loop can be implemented using

$$u_0(t) = k_1 \cdot r(t) - (k_1 \cdots k_N) \cdot (\hat{x}_1(t) \cdots \hat{x}_N(t))^{\mathrm{T}}. \qquad (3.14)$$

Putting (3.14) into (3.13) yields the combined, overall control law of ADRC:

$$u(t) = \frac{1}{b_0} \cdot \left(k_1 \cdot r(t) - \left(\boldsymbol{k}^{\mathrm{T}} \; 1 \right) \cdot \hat{\boldsymbol{x}}(t) \right) \qquad (3.15)$$

$$\text{with} \quad \boldsymbol{k}^{\mathrm{T}} = \left(k_1 \cdots k_N \right) \quad \text{and} \quad \hat{\boldsymbol{x}} = \left(\hat{x}_1 \cdots \hat{x}_{N+1} \right)^{\mathrm{T}}.$$

In (3.15), b_0 is the critical gain parameter to be obtained during plant modeling and $\boldsymbol{k}^{\mathrm{T}}$ the vector of state-feedback gains that have to be tuned to achieve desired closed-loop dynamics. The state vector $\hat{\boldsymbol{x}}(t)$ is being estimated by a Luenberger observer with a feedback gain vector $\boldsymbol{l}$, to be tuned for faster response than the outer loop:

$$\dot{\hat{\boldsymbol{x}}}(t) = \boldsymbol{A} \cdot \hat{\boldsymbol{x}}(t) + \boldsymbol{b} \cdot u(t) + \boldsymbol{l} \cdot \left(y(t) - \boldsymbol{c}^{\mathrm{T}} \cdot \hat{\boldsymbol{x}}(t) \right) \qquad (3.16)$$

$$\text{with} \quad \boldsymbol{A} = \begin{pmatrix} \boldsymbol{0}^{N \times 1} & \boldsymbol{I}^{N \times N} \\ 0 & \boldsymbol{0}^{1 \times N} \end{pmatrix}, \; \boldsymbol{b} = \begin{pmatrix} \boldsymbol{0}^{(N-1) \times 1} \\ b_0 \\ 0 \end{pmatrix}, \; \boldsymbol{l} = \left(l_1 \cdots l_{N+1} \right)^{\mathrm{T}}, \; \boldsymbol{c}^{\mathrm{T}} = \left(1 \; \boldsymbol{0}^{1 \times N} \right).$$

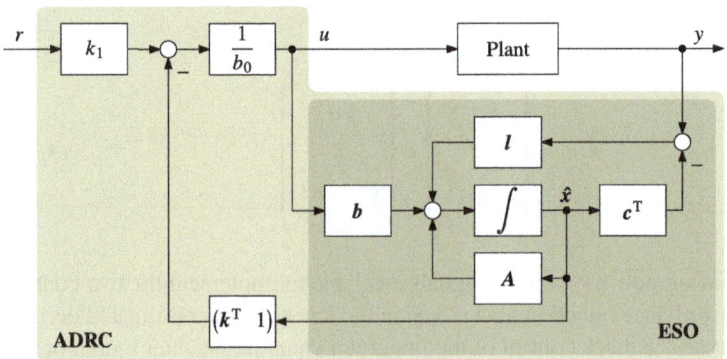

Fig. 3.5 Block diagram of a control loop with continuous-time linear ADRC in its state-space form

Figure 3.5 summarizes all equations of ADRC in block diagram form. Being based on the combined control law (3.15), the two control loops have apparently been merged. Nonetheless, they are present, and it is still possible to recognize the two core ideas of the inner loop: *plant gain inversion* (implemented using the gain block $\frac{1}{b_0}$) and *disturbance rejection*. The details of the latter are a bit hidden in the vector form of signals and gains but surface when analyzing the two instruments from linear control theory employed in ADRC:

- *Extended state observer (ESO):* For linear ADRC of order N, an observer as given in (3.9) estimates a filtered version of the plant output y and its first $(N-1)$ derivatives, extended by an additional state $\hat{x}_{N+1}$ representing the *total disturbance*. Due to this extension, in the context of ADRC, the name *extended state observer* is commonly being used, but be assured that the tool employed here is the well-known Luenberger observer—it is just about the particular choice of the model inside.

  Within the observer, the *total disturbance* is being modeled as a constant input disturbance to a plant with integrator chain dynamics. Note that a constant input disturbance represents the most common case in practice and the one corresponding to the disturbance rejection abilities of PID controllers. If required by the application, more sophisticated disturbance models can be incorporated into the observer, as well.[2]

- *State feedback:* A linear feedback of the states $\hat{x}$ combines inner and outer control loops: firstly, a state-feedback controller with gain k^T for the integrator chain plant constituted by the inner loop (states $\hat{x}_1, \ldots, \hat{x}_N$ estimated by the observer), and secondly, a unity-gain feedback of the state $\hat{x}_{N+1}$, represented by the factor 1 as part of the feedback gain vector $(k^T \ 1)$. The latter implements rejection of the *total disturbance* in the inner loop. Finally, the output of the state-feedback controller is multiplied by a factor $\frac{1}{b_0}$, the reciprocal value of the plant gain b_0. Creating a unity-gain plant for the outer state-feedback controller considerably eases its tuning, as discussed in Sect. 3.1.3 and once more to be seen in Sect. 3.2.1.

Controller and observer in (3.15) and (3.16) are here given for the general, Nth-order case. In practice, $N = 1$ and $N = 2$ are most relevant, corresponding to the use cases of PI and PID controllers. For those, Table A.1.2 on page 192 of the Appendix specifies A and b in unabbreviated form.

3.2 Parameter Tuning

Let us briefly recapitulate parameters that need to be selected or tuned for ADRC. Plant modeling involves the plant gain b_0 and order N—refer to Table 3.1 for examples. The latter will influence the number of observer state variables $(N+1)$ and hence the number of feedback gains that must be tuned. We will now address computing those N controller gains in k^T and $(N+1)$ observer gains in l.

---
[2] We will shed some more light on this in Sect. 4.1 and provide an example in Sect. 6.1.

3.2.1 Tuning the Controller

The combined control law (3.15) consists of disturbance rejection, plant gain inversion, and the output of an outer linear state-feedback controller. As discussed in Sect. 3.1.2, the disturbance-compensating feedback of $\hat{x}_{N+1}(t)$ can, in conjunction with plant gain inversion, be viewed as an inner control loop. As depicted in Fig. 3.6, this creates a "virtual plant," which is then to be controlled by the (outer) state-feedback controller (3.14).

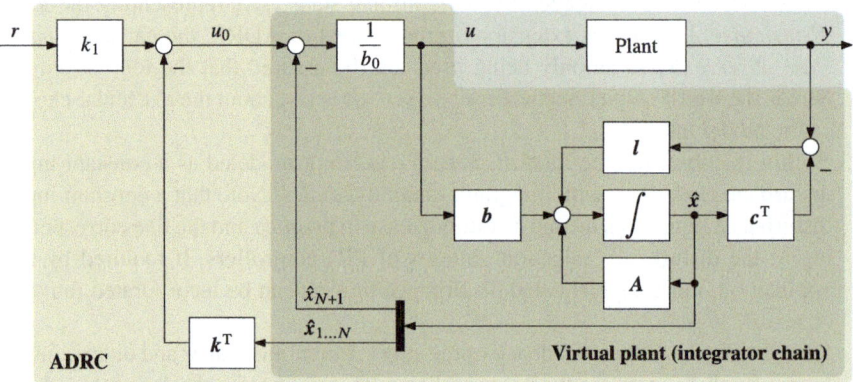

Fig. 3.6 Control loop with continuous-time ADRC, where the compensation of the *total disturbance* is seen as an inner control loop with the actual plant. This inner loop forms a virtual plant to the outer state-feedback controller k^T and has a unity-gain integrator chain behavior

If we assume the observer to be fast enough, rejecting the *total disturbance* will be so effective that the state-feedback controller appears to control a plant with unity-gain integrator chain behavior—regardless of the actual plant. This is the result of "normalizing" the plant behavior in Sect. 3.1.2, which makes controller tuning very easy. A state-space representation of such a virtual plant $y^{(N)}(t) = u_0(t)$ is

$$\dot{x}_{\text{VP}}(t) = A_{\text{VP}} \cdot x_{\text{VP}}(t) + b_{\text{VP}} \cdot u_0(t), \quad y(t) = c_{\text{VP}}^T \cdot x_{\text{VP}}(t) \quad (3.17)$$

with $A_{\text{VP}} = \begin{pmatrix} \mathbf{0}^{(N-1)\times 1} & I^{(N-1)\times(N-1)} \\ 0 & \mathbf{0}^{1\times(N-1)} \end{pmatrix}$, $b_{\text{VP}} = \begin{pmatrix} \mathbf{0}^{(N-1)\times 1} \\ 1 \end{pmatrix}$, $c_{\text{VP}}^T = \begin{pmatrix} 1 & \mathbf{0}^{1\times(N-1)} \end{pmatrix}$,

in which the state variables are defined to match the meaning of the first N state variables in the observer model of (3.9): $x_{\text{VP},1}(t)$ equals the output $y(t)$, and $x_{\text{VP},i+1}(t)$ is the ith derivative of $y(t)$ for $i = 1, \ldots, N-1$.

A possible and common approach to shape the closed-loop behavior is pole placement, i.e., computing the N gains of the vector k^T such that the system matrix of the closed loop $(A_{\text{VP}} - b_{\text{VP}} k^T)$ features N eigenvalues as desired: the poles of the closed-loop transfer function.

3.2 Parameter Tuning

The freedom of selecting N eigenvalues may, however, be experienced as a burden by many users. A practical approach greatly simplifying the controller tuning process is known under the name *bandwidth parameterization* and consists of pole placement with all poles at a common location $\lambda = -\omega_{\text{CL}}$, the desired closed-loop bandwidth. The characteristic polynomial of the closed-loop system matrix then becomes

$$(\lambda + \omega_{\text{CL}})^N \stackrel{!}{=} \det\left(\lambda \boldsymbol{I} - \left(\boldsymbol{A}_{\text{VP}} - \boldsymbol{b}_{\text{VP}} \boldsymbol{k}^{\text{T}}\right)\right) = \lambda^N + k_N \lambda^{N-1} + \ldots + k_2 \lambda + k_1. \quad (3.18)$$

After applying binomial expansion to the left-hand side of (3.18), the controller gains k_i can be read off as follows:

$$k_i = \frac{N!}{(N-i+1)! \cdot (i-1)!} \cdot \omega_{\text{CL}}^{N-i+1} \quad \forall i = 1, \ldots, N. \quad (3.19)$$

While the generic expression for an Nth-order system in (3.19) might appear a bit convoluted, very simple tuning rules relating the desired bandwidth ω_{CL} to the k_i gains emerge for the two cases most relevant in practice:

- For $N = 1$, $k_1 = \omega_{\text{CL}}$.
- For $N = 2$, $k_1 = \omega_{\text{CL}}^2$ and $k_2 = 2\omega_{\text{CL}}$.

Example 3.1. To demonstrate the use of (3.19), controller gains shall be computed for second-order ADRC ($N = 2$) with a desired closed-loop bandwidth of 0.2 Hz, i.e., $\omega_{\text{CL}} = 2\pi \cdot 0.2 \,\text{rad/s}$. Solution: $k_1 = \omega_{\text{CL}}^2 \approx 1.579 \, \frac{1}{s^2}$ and $k_2 = 2\omega_{\text{CL}} \approx 2.513 \, \frac{1}{s}$.

Frequency-Domain Derivation

For readers unfamiliar with the process of placing eigenvalues in a state-feedback control system, one can arrive at the same result in the frequency domain. Taking the Laplace transform of the (idealized) closed-loop behavior (3.7) and comparing that with a critically damped Nth-order low-pass filter (only real poles) yield

$$G_{\text{CL}}(s) = \frac{y(s)}{r(s)} = \frac{1}{\frac{1}{k_1} s^N + \frac{k_N}{k_1} \cdot s^{N-1} + \ldots + \frac{k_2}{k_1} \cdot s + 1} \stackrel{!}{=} \frac{1}{\left(\frac{s}{\omega_{\text{CL}}} + 1\right)^N},$$

where, after multiplying both sides with $k_1 = \omega_{\text{CL}}^N$, we can compare the denominators more easily:

$$s^N + k_N \cdot s^{N-1} + \ldots + k_2 \cdot s + k_1 \stackrel{!}{=} (s + \omega_{\text{CL}})^N. \quad (3.20)$$

Obviously (3.20) has the same structure as (3.18), which means one will obtain the same results for the controller gains k_i as given in (3.19), e.g., $k_1 = \omega_{\text{CL}}^N$ and $k_N = N \cdot \omega_{\text{CL}}$.

Tuning Based on Time-Domain Characteristics

A time-domain interpretation may simplify choosing the frequency-domain tuning goal ω_{CL}. A common measure in the time domain is the settling time of a control loop, where the controlled variable reaches a certain percentage of the final value, subsequently further converging. Various approximations balancing simplicity against accuracy can be found, such as $\omega_{\text{CL}} \approx \frac{4}{T_{\text{settle},98\%}}$ and $\omega_{\text{CL}} \approx \frac{6}{T_{\text{settle},98\%}}$ for the first- and second-order cases, respectively. With a more accurate approximation, the following relation provides an ω_{CL} value leading to the desired 98 % settling time with less than 1 % error for orders from $N = 1$ to $N = 6$, as shown in Fig. 3.7:

$$\omega_{\text{CL}} \approx \frac{1.9 + 2.09N - 0.07N^2}{T_{\text{settle},98\%}}. \tag{3.21}$$

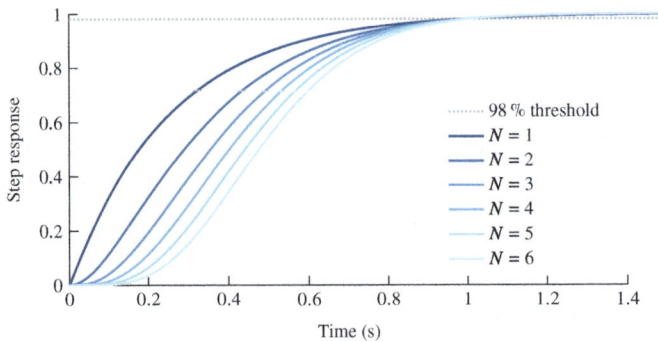

Fig. 3.7 Step responses of systems of order $N = 1$ to $N = 6$ tuned for a 98 % settling time of $T_{\text{settle},98\%} = 1$ s using *bandwidth parameterization* in conjunction with the approximation (3.21)

3.2.2 Tuning the Observer

Computing the feedback gain vector l of ESO will conclude the tuning procedure required for linear ADRC. As with the controller gain vector k^T, there is freedom to choose any possible design approach for l. A pragmatic choice is to employ pole placement with *bandwidth parameterization* also for the observer, again to lower the overall number of tuning parameters.

The *total disturbance* must be estimated so fast that the inner loop of Fig. 3.6 ("virtual plant") maintains the desired integrator chain behavior. Also, the derivatives of the plant output must be provided by the observer fast enough to serve as a meaningful basis for the outer state-feedback controller. This is, as usual in observer-based control, only possible if the observer bandwidth ω_O exceeds the desired

3.3 Summary and Outlook

closed-loop bandwidth ω_{CL}. Therefore, in this book, we will express the observer bandwidth depending on the closed-loop bandwidth using a relative factor k_{ESO}:

$$\omega_{\text{O}} = k_{\text{ESO}} \cdot \omega_{\text{CL}}. \tag{3.22}$$

Using k_{ESO} shall emphasize that the bandwidths cannot be independently selected. As a rule of thumb, the observer should be tuned for a three- to tenfold bandwidth compared to ω_{CL}. In practice, an upper bound will be often determined by the acceptable noise level of the control signal. A lower bound is formed when closed-loop dynamics significantly deviate from the intended design, i.e., when the observer is too slow to fully reject disturbances and maintain the "virtual plant" behavior.

The characteristic polynomial of the observer system matrix $(A - lc^{\text{T}})$ in the case of bandwidth parameterization for a common pole location $\lambda = -\omega_{\text{O}} = -k_{\text{ESO}} \cdot \omega_{\text{CL}}$ has to satisfy the following equation:

$$(\lambda + k_{\text{ESO}} \cdot \omega_{\text{CL}})^{N+1} \stackrel{!}{=} \det\left(\lambda I - \left(A - lc^{\text{T}}\right)\right) = \lambda^{N+1} + l_1 \lambda^N + \ldots + l_N \lambda + l_{N+1}. \tag{3.23}$$

The similarity of (3.24) and (3.18) is not accidental given the structural similarities of the A and A_{VP} matrices and the b and c^{T} vectors. Binominal expansion of the left-hand side of (3.23) yields the general equation for computing ADRC observer gains under the *bandwidth parameterization* approach:

$$l_i = \frac{(N+1)!}{(N-i+1)! \cdot i!} \cdot (k_{\text{ESO}} \cdot \omega_{\text{CL}})^i \quad \forall i = 1, \ldots, N+1. \tag{3.24}$$

Particularly relevant in practice are values for the first- and second-order cases:

- For $N = 1$, $l_1 = 2k_{\text{ESO}}\omega_{\text{CL}}$ and $l_2 = k_{\text{ESO}}^2 \omega_{\text{CL}}^2$.
- For $N = 2$, $l_1 = 3k_{\text{ESO}}\omega_{\text{CL}}$, $l_2 = 3k_{\text{ESO}}^2 \omega_{\text{CL}}^2$, and $l_3 = k_{\text{ESO}}^3 \omega_{\text{CL}}^3$.

Example 3.2. Continuing example 3.1, the observer gains shall now be computed, with an observer bandwidth that is $k_{\text{ESO}} = 5$ times higher than the closed-loop bandwidth ω_{CL}. Solution: $l_1 = 3k_{\text{ESO}}\omega_{\text{CL}} \approx 18.85 \frac{1}{\text{s}}$, $l_2 = 3k_{\text{ESO}}^2 \omega_{\text{CL}}^2 \approx 118.4 \frac{1}{\text{s}^2}$, $l_3 = k_{\text{ESO}}^3 \omega_{\text{CL}}^3 \approx 248.1 \frac{1}{\text{s}^3}$. Hint: If you cannot wait to see this controller/observer pair in action, have a peek at Sect. 5.2, where we present a first closed-loop simulation example using exactly these parameters.

3.3 Summary and Outlook

Whenever we introduce a new variant of ADRC in this book that can be put into practice—be it in the form of a simulation model or later as a custom software implementation—we will summarize the required steps in the form of a "cooking recipe." We have now reached such a point for the first time.

Cooking Recipe (ADRC, State-Space Form)

To implement and tune linear active disturbance rejection control in its continuous-time state-space form, these steps have to be carried out:

1. *Plant modeling:* For the dominant dynamic plant behavior, identify the order N and the parameter b_0 of the plant model (3.2). Refer to Table 3.1 for some simple examples.
2a. *Controller tuning:* ADRC of order N requires tuning of N controller gains $k_{1,...,N}$. When using *bandwidth parameterization* for the desired closed-loop dynamics, these gain values are obtained from (3.19) or Table A.1.1. If necessary, use an approximation as (3.21) to translate a time-domain specification into the tuning parameter ω_{CL}.
2b. *Observer tuning:* ADRC of order N requires $(N+1)$ observer gains $l_{1,...,N+1}$. In the case of *bandwidth parameterization*, choosing a relative observer bandwidth $k_{\text{ESO}} = 3, \ldots, 10$ is a good starting point. Use (3.24) or Table A.1.1 to compute the gains from k_{ESO} and ω_{CL}.
3. *Implementation:* Implement the state-space form of ADRC as shown in Fig. 3.5. Required matrices and vectors are defined in the equations of the controller (3.15) and the extended state observer (3.16).

To serve as a quick reference guide, all relevant equations of this chapter are once more compiled in Appendix A.1. We will stick to this pattern throughout this book: "cooking recipes" summarizing a new variant whenever introduced, a detailed list of "ingredients" in the Appendix. Please bear in mind that a real-world controller contains more than the bare control law. Most notably a limitation of the controller output is still missing here, along with some form of anti-windup protection. On our way to a practical implementation, we will come back to this repeatedly in this book.

Regarding the nomenclature, note again that we speak of Nth-order ADRC if the assumed plant model (3.2) is of order N, which means there are N controller gains, $(N-1)$ estimated derivatives, and at least $(N+1)$ state variables in the observer: output, derivatives, and disturbance model.

Finally, let us stress once more that we have adopted the *linear* variant of ADRC as the basis for this book given its several advantages:

- Methods: One can build on a century of linear control systems theory, e.g., concerning tuning methods or stability analyses.
- Recognition factor: Connections and similarities to the existing, "classical" control world can be seen more easily.
- Entry barrier: Being well-rooted in "proven technology" helps to lower the barriers to adopting ADRC, be it from an engineering or a management perspective.

For brevity, we will often refer to this linear variant (and its various descendants and forms that will be introduced in the book) simply as "ADRC"—the acronym that we also use to describe the whole field. Whenever a distinction needs to be made or the property of linearity needs to be highlighted, we will use the acronym "LADRC."

3.3 Summary and Outlook

Next Steps: Deeper Understanding

To facilitate further reading, in the upcoming three segments we offer the reader guidance through the remainder of this book. The first set of reading options revolves around deepening the understanding of ADRC:

- Chapter 4 discusses linear ADRC from a classical control engineering perspective. Firstly, the connection to observer-based state-space control as known from linear systems theory is established in Sect. 4.1. For readers with a preference for frequency rather than time-domain tools, Sect. 4.2 derives transfer function representations of linear ADRC.
- Chapter 5 continues with a comprehensive, visual-heavy analysis in both time and frequency domains, including the following aspects: How to interpret or obtain the critical gain parameter of a plant (Sect. 5.1)? What is the role of the estimated state variables, notably the *total disturbance*, in a typical control scenario (Sect. 5.2)? What is the influence of the "tuning knobs" (bandwidth, observer bandwidth, and critical gain parameter) on tracking performance, disturbance rejection abilities, and stability (Sect. 5.3)? Part of Sect. 5.3.3 will also be a frequency-domain comparison to the transfer functions of PI and PID controllers. Finally, ADRC's ability to cope with parametric or structural uncertainties of the plant to be controlled is examined in Sect. 5.4.
- Chapter 7 will make you familiar with relevant literature from the field of ADRC.

Next Steps: Implementation

The intention of this book is to guide from principles to practical implementation. In Chap. 8, it therefore works its way up to ready-to-use discrete-time forms of linear ADRC suitable for an (usually software-based) implementation.

- A discrete-time form in state space is derived in Sect. 8.1. This is the first ADRC variant that can actually be implemented in software and the basis for all other discrete-time implementations covered in this book.
- Based on that, a discrete-time transfer function form is introduced in Sect. 8.2. This form is very similar to classical digital controllers but—as will be seen—does not retain all features from the state-space variant.
- A second discrete-time transfer function form with two feedback loops is introduced in Sect. 8.3, keeping all features from the state-space variant. It can be implemented with a very low computational footprint, thus combining the advantages of the state space and transfer function worlds.

Depending on the reader's previous experience, descending from an "equation level" view of ADRC to the lowlands of an actual (software) implementation might or might not appear easy. As this topic is (unfortunately, in our view) not always part of control engineering curricula, we have dedicated Chap. 10 to support this transition in both model-based (Sect. 10.1) and handwritten source code implementations (Sect. 10.2). Our application examples in Chap. 11 will make use of these.

Next Steps: Customization

It has to be stressed that, as presented in this chapter, you have just seen *one* form of ADRC. But please do not view ADRC as an immutable set of equations, but rather as a certain way of looking at, understanding, and solving control problems. From the core that we chose with linear ADRC, further variants and implementations can be derived. At the moment, many variations and extensions exist and can affect all ingredients. The following ones are covered in this book:

- Incorporation of additional plant model information in the observer, deviating from the simple integrator chain model, again to improve the dynamic performance. We will examine in Sect. 6.1 if and how additional information about the plant or disturbance improves tracking behavior and disturbance rejection.
- In Sect. 6.2, we will examine if and how additional information about the reference signal derivatives improves the tracking behavior.
- If needed, some performance aspects of ADRC can potentially be improved by adding nonlinearity to its structure. Different tools and methods for improving tracking and estimation quality utilizing nonlinear ADRC schemes are discussed in Sect. 6.3.
- Error-based ADRC allows obtaining implementations even more similar to classical control engineering and is introduced in Sect. 6.4. It can particularly well be combined with arbitrary reference signal prefilters, allowing accomplishing true two-degrees-of-freedom behavior. The implementation of discrete-time error-based ADRC can be found in Sect. 8.4.
- Further practical issues such as dealing with control signal limitations, smooth controller enabling, measurement noise, or process dead time are discussed in Chap. 9.

As you see, nothing is set in stone. Yet we have made specific choices in this book for good reasons, trying to guide you from principles to practice in a consistent manner. You can view linear ADRC from this chapter as a sound, practically proven starting point—a foundation you can build on and customize to your application's needs.

Open Access This chapter is licensed under the terms of the Creative Commons Attribution 4.0 International License (http://creativecommons.org/licenses/by/4.0/), which permits use, sharing, adaptation, distribution and reproduction in any medium or format, as long as you give appropriate credit to the original author(s) and the source, provide a link to the Creative Commons license and indicate if changes were made.

The images or other third party material in this chapter are included in the chapter's Creative Commons license, unless indicated otherwise in a credit line to the material. If material is not included in the chapter's Creative Commons license and your intended use is not permitted by statutory regulation or exceeds the permitted use, you will need to obtain permission directly from the copyright holder.

Chapter 4
Between Time and Frequency Domains

Abstract In this chapter, we will contextualize ADRC to deliver interpretations from the established field of linear control systems. This will, firstly, involve an interpretation of ADRC in terms of existing state-space control approaches. Secondly, a transition to the frequency domain is being made, as a major part of practical control engineering is performed in this domain. Analyzing and understanding ADRC from a frequency-domain perspective allows for a much easier migration from the PID world of controllers.

4.1 ADRC as a State-Space Controller

We have seen in Chap. 3 that the ADRC variant forming the core of all controllers in this book only consists of tools well known from linear systems theory: observer, state feedback, and disturbance compensation. In this section, we therefore want to precisely reveal the connection to "classical" state-space control.

State-Feedback Control with Disturbance Compensation

To introduce the "textbook" approach of observer-based state-feedback control with disturbance compensation, consider a state-space description of a plant with output y, input u, and an input disturbance d, all being scalar values:

$$\begin{aligned}\dot{\boldsymbol{x}}(t) &= \boldsymbol{A} \cdot \boldsymbol{x}(t) + \boldsymbol{b} \cdot u(t) + \boldsymbol{e} \cdot d(t), \\ y(t) &= \boldsymbol{c}^{\mathrm{T}} \cdot \boldsymbol{x}(t).\end{aligned} \quad (4.1)$$

Let us assume that the dynamics behind the nonmeasurable disturbance $d(t)$ can be described by the following equation[1]:

$$\dot{x}_d(t) = A_d \cdot x_d(t),$$
$$d(t) = c_d^T \cdot x_d(t). \tag{4.2}$$

These two equations can be combined to an extended process description:

$$\begin{pmatrix} \dot{x}(t) \\ \dot{x}_d(t) \end{pmatrix} = \begin{pmatrix} A & e \cdot c_d^T \\ 0 & A_d \end{pmatrix} \cdot \begin{pmatrix} x(t) \\ x_d(t) \end{pmatrix} + \begin{pmatrix} b \\ 0 \end{pmatrix} \cdot u(t),$$
$$y(t) = (c^T \ 0) \cdot \begin{pmatrix} x(t) \\ x_d(t) \end{pmatrix}. \tag{4.3}$$

Assuming that only the plant output $y(t)$ is measurable, a full-order Luenberger observer for the process model augmented by the disturbance model is now being set up, in order to serve as the basis for state-feedback control:

$$\begin{pmatrix} \dot{\hat{x}}(t) \\ \dot{\hat{x}}_d(t) \end{pmatrix} = \underbrace{\begin{pmatrix} A & e \cdot c_d^T \\ 0 & A_d \end{pmatrix}}_{A_o} \cdot \begin{pmatrix} \hat{x}(t) \\ \hat{x}_d(t) \end{pmatrix} + \underbrace{\begin{pmatrix} b \\ 0 \end{pmatrix}}_{b_o} \cdot u(t) + k_o \cdot \left(y(t) - \underbrace{(c^T \ 0)}_{c_o^T} \cdot \begin{pmatrix} \hat{x}(t) \\ \hat{x}_d(t) \end{pmatrix} \right). \tag{4.4}$$

The feedback gain inside the observer (4.4) is k_o. Using the estimated states of the plant $\hat{x}(t)$ and the estimated states of the disturbance generator model $\hat{x}_d(t)$, a state-feedback control law can now be implemented as follows:

$$u(t) = k_s \cdot r(t) - k_c^T \cdot \hat{x}(t) - k_d^T \cdot \hat{x}_d(t). \tag{4.5}$$

The overall control loop is shown in Fig. 4.1, with the controller (4.5) comprising the following elements:

- *State feedback:* Feeding back the estimated states $\hat{x}(t)$ with the gain vector k_c^T moves the poles of the plant as desired for the closed-loop dynamics, which are determined by the eigenvalues of $(A - bk_c^T)$.
- *Static gain compensation:* To ensure $r = y$ in steady state, the gain k_s must be set to the inverse of the loop gain using (4.6):

$$k_s = -\left[c^T \cdot \left(A - bk_c^T \right)^{-1} \cdot b \right]^{-1}. \tag{4.6}$$

- *Disturbance compensation:* Feedback of $\hat{x}_d(t)$ via k_d^T to compensate the influence of the disturbance $d(t)$ on the plant. Inserting (4.5) in (4.1) reveals that the disturbance may be fully rejected if accurately estimated and if $b \cdot k_d^T = e \cdot c_d^T$ can be achieved.

---

[1] The subscript "d" in (4.2) refers to the disturbance d in this section. Later in this book, in Sect. 8.1, this subscript will be reused for discrete-time values; please do not mistake them.

4.1 ADRC as a State-Space Controller

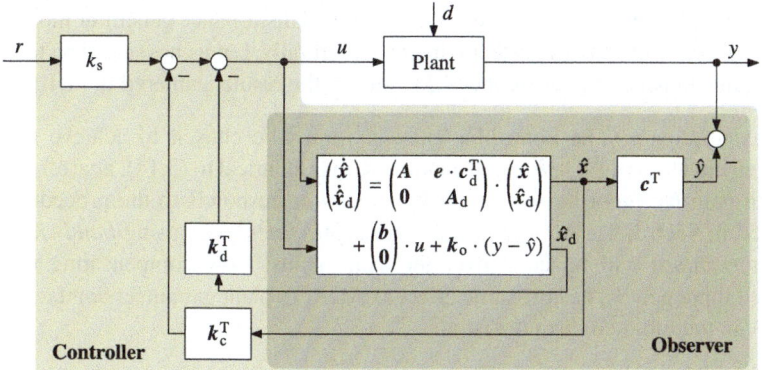

Fig. 4.1 Generic structure of an observer-based state-feedback control loop with scalar input/output signals. The observer is augmented with a model for the disturbance d. State feedback consists of one path for disturbance compensation with k_d and one path for feeding back the estimated states of the plant using k_c

Comparison with Linear ADRC

We have now gathered all the facts required to perform a detailed comparison of ADRC with the generic approach of observer-based state-feedback control with disturbance compensation.

- *Disturbance generator model:* The disturbance model of ADRC in (3.9) is an unknown constant value for the *total disturbance* $f(t)$; therefore $\hat{x}_\text{d}(t)$ only consists of one scalar value. If we consequently set $d(t) = f(t)$ and $\hat{x}_\text{d}(t) = \hat{x}_\text{d}(t) = \hat{f}(t)$, we obtain for the disturbance generator model (4.2)

$$A_\text{d} = \begin{pmatrix} 0 \end{pmatrix}, \quad c_\text{d}^\text{T} = \begin{pmatrix} 1 \end{pmatrix}. \tag{4.7}$$

- *Plant model:* Comparing (4.4) with the corresponding elements in (3.9) gives the values for A, b, c^T of the plant (4.1):

$$A = \begin{pmatrix} \mathbf{0}^{(N-1)\times 1} & \mathbf{I}^{(N-1)\times(N-1)} \\ 0 & \mathbf{0}^{1\times(N-1)} \end{pmatrix}, \quad b = \begin{pmatrix} \mathbf{0}^{(N-1)\times 1} \\ b_0 \end{pmatrix}, \quad c^\text{T} = \begin{pmatrix} 1 & \mathbf{0}^{1\times(N-1)} \end{pmatrix}. \tag{4.8}$$

Additionally, using the values from (4.7), we can read off e from (4.4):

$$e = \begin{pmatrix} \mathbf{0}^{(N-1)\times 1} \\ 1 \end{pmatrix}. \tag{4.9}$$

- *Control law:* Comparing the control law of ADRC from (3.15) with (4.5) gives

$$k_s = \frac{k_1}{b_0}, \quad k_\text{c}^\text{T} = \frac{1}{b_0} \cdot k^\text{T} = \frac{1}{b_0} \cdot \begin{pmatrix} k_1 & \cdots & k_N \end{pmatrix}^\text{T}, \quad k_\text{d}^\text{T} = k_\text{d} = \frac{1}{b_0}. \tag{4.10}$$

So far we could determine the values of the plant and disturbance generator model and have seen that control laws perfectly match structurally. Let us now confirm that tuning the gains in generic control law (4.5) delivers the results gathered in (4.10).

- *State-feedback gain:* The controller gains k_c^T must be chosen to achieve the desired eigenvalues of $(A - bk_c^T)$. Equation (4.8) is identical to (3.17), apart from the input gain b_0. Ignoring b_0, tuning k_c^T exactly corresponds to the procedure described in Sect. 3.2.1. Aiming at the same eigenvalues (e.g., using *bandwidth parameterization*) will hence deliver the same gains. Now compensating the neglected input gain b_0 has the same effect as ADRC's plant gain inversion factor $\frac{1}{b_0}$. We can therefore confirm the result:

$$k_c^T = \frac{1}{b_0} \cdot k^T = \frac{1}{b_0} \cdot (k_1 \cdots k_N). \tag{4.11}$$

- *Static gain compensation:* Putting (4.8) and (4.11) into (4.6) gives the necessary gain k_s to compensate the inner loop gain:

$$k_s = -\left[\left(1 \ \mathbf{0}^{1\times(N-1)}\right) \cdot \left(\begin{pmatrix} \mathbf{0}^{(N-1)\times 1} & I^{(N-1)\times(N-1)} \\ 0 & \mathbf{0}^{1\times(N-1)} \end{pmatrix} \right. \right.$$
$$\left. \left. - \begin{pmatrix} \mathbf{0}^{(N-1)\times 1} \\ b_0 \end{pmatrix} \cdot \frac{1}{b_0} \cdot (k_1 \cdots k_N)^T \right)^{-1} \cdot \begin{pmatrix} \mathbf{0}^{(N-1)\times 1} \\ b_0 \end{pmatrix} \right]^{-1}.$$

Simplifying this equation leads to the confirmation of the result from (4.10):

$$k_s = \frac{k_1}{b_0}. \tag{4.12}$$

- *Disturbance compensation gain:* As mentioned above, if an accurate estimation of the states $\hat{x}_d(t)$ is provided by the observer, the disturbance $d(t)$ can be fully compensated if the following condition is satisfied:

$$b \cdot k_d^T = e \cdot c_d^T,$$

i.e., $\begin{pmatrix} \mathbf{0}^{(N-1)\times 1} \\ b_0 \end{pmatrix} \cdot k_d = \begin{pmatrix} \mathbf{0}^{(N-1)\times 1} \\ 1 \end{pmatrix} \cdot 1.$

We can therefore confirm the value obtained by comparison in (4.10):

$$k_d = \frac{1}{b_0}. \tag{4.13}$$

As we have, using (4.7) and (4.8), established full equivalence also for the augmented plant model in the observer (4.4), it is clear that tuning the observer gains k_o with the same approach (e.g., *bandwidth parameterization*) will yield the same results as ADRC's observer tuning described in Sect. 3.2.2, i.e., $k_o = l$ will hold.

Discussion

We have now traced linear ADRC back to "classical" observer-based state-feedback control with disturbance compensation and established full structural equivalence. What does that mean? Is ADRC nothing new and just constituted by a certain parameter choice?

Yes and no. Being structurally equivalent to well-known approaches from linear control systems means that all methods to analyze stability and performance can be applied to LADRC, as well, which is good news. Pointing out that LADRC boils down to a parameter choice, however, misses its point. This would correspond to claiming that PID controllers are just a parameter choice of transfer functions (in Laplace domain). While technically true, the major philosophical difference of ADRC is *how* these parameters are obtained. In classical state-feedback control, the observer is equipped with a plant model as accurately as possible. For ADRC, however, one deliberately makes the (erroneous) assumption of an integrator chain model, regardless of the actual plant behavior. This greatly simplifies plant modeling and represents a departure from the established model-based control school.[2]

As a final note we also want to mention that, for ADRC, the disturbance generator model does not always have to be as simple as a constant disturbance. However, this is a very reasonable default choice, which structurally corresponds to the disturbance rejection abilities of any classical controller with an integral component (such as PI, PID, or pure I controllers). More sophisticated disturbance models are possible, and some examples will be discussed in Chap. 7. This could, for instance, be needed when trying to fully reject a sinusoidal disturbance. The insights provided by this chapter should serve as a starting point to customize ADRC with an application-tailored disturbance generator model.

4.2 Transfer Function Representation

Previously in Sect. 4.1 we have seen how linear ADRC is rooted in the established world of state-space control systems. As we have positioned ADRC as a practical tool aiming at sprawling in the territory of PI and PID controllers, we want to provide a means of understanding ADRC from a classical control engineering point of view, which usually involves treatment in the frequency domain. To this end, a transfer function representation will be derived here. And, as will be shown in Sect. 4.2.2, ADRC can—with a little extra effort—even be represented using *realizable* transfer functions. A comparison of these transfer functions to existing controllers enables deeper insights and understanding of ADRC, especially for control engineers native to the frequency domain world. And, not least, a transfer function representation will pave the way for efficiently implementing ADRC—something we will later discuss in detail for the discrete-time case in Chap. 8.

---

[2] In the historical and literature overview on ADRC delivered in Chap. 7, we will point out that this very fact is considered a true "paradigm shift" and of ADRC's key selling points.

Deriving the transfer functions begins with the closed-loop observer dynamics. The Laplace transform of the control law (3.15) is put in the observer (3.16):

$$\hat{x}(s) = \left(sI - \underbrace{\left(A - lc^T - \frac{1}{b_0}b\left(k^T\ 1\right)\right)}_{A_{CL}}\right)^{-1} \cdot \left(\frac{k_1}{b_0}br(s) + ly(s)\right). \quad (4.14)$$

To shorten further equations, we will use the abbreviated system matrix A_{CL} of the closed-loop controller/observer dynamics as introduced in (4.14):

$$A_{CL} = A - lc^T - \frac{1}{b_0}b\left(k^T\ 1\right) = \begin{pmatrix} \mathbf{0}^{N\times 1} & I^{N\times N} \\ 0 & \mathbf{0}^{1\times N} \end{pmatrix} - \left(l\ \mathbf{0}^{(N+1)\times N}\right) - \begin{pmatrix} \mathbf{0}^{(N-1)\times N} & 0 \\ k^T & 1 \\ \mathbf{0}^{1\times N} & 0 \end{pmatrix}. \quad (4.15)$$

Presenting (4.15) in full detail reveals that, while being an $(N+1)\times(N+1)$ matrix, the rank of A_{CL} cannot exceed N, as its rightmost column is always zero:

$$A_{CL} = \begin{pmatrix} -l_1 & 1 & 0 & \cdots & 0 & 0 & 0 \\ -l_2 & 0 & 1 & \cdots & 0 & 0 & 0 \\ -l_3 & 0 & 0 & \cdots & 0 & 0 & 0 \\ \vdots & \vdots & \ddots & \ddots & \ddots & \vdots \\ -l_{N-1} & 0 & 0 & \cdots & 0 & 1 & 0 \\ -l_N - k_1 & -k_2 & -k_3 & \cdots & -k_{N-1} & -k_N & 0 \\ -l_{N+1} & 0 & 0 & \cdots & 0 & 0 & 0 \end{pmatrix}. \quad (4.16)$$

A transfer function of the controller can now be obtained by putting (4.14) back in the Laplace transform of (3.15), eliminating $\hat{x}(s)$ from the equations:

$$u(s) = \frac{k_1}{b_0}r(s) - \frac{1}{b_0}\left(k^T\ 1\right)\cdot(sI - A_{CL})^{-1}\cdot\left(\frac{k_1}{b_0}br(s) + ly(s)\right). \quad (4.17)$$

This can be reformulated with a common denominator polynomial:

$$u(s) = \left(b_0 \cdot \det(sI - A_{CL})\right)^{-1} \cdot$$
$$\left[\left(k_1 \cdot \det(sI - A_{CL}) - \frac{k_1}{b_0}\cdot(k^T\ 1)\cdot\text{adj}(sI - A_{CL})\cdot b\right)\cdot r(s)\right.$$
$$\left. - \left((k^T\ 1)\cdot\text{adj}(sI - A_{CL})\cdot l\right)\cdot y(s)\right]. \quad (4.18)$$

Equation (4.18), which relates the two inputs (r and y) of ADRC's control law to its output (u) in Laplace domain, can now serve as the common starting point for the two transfer function representations introduced next.

4.2 Transfer Function Representation

4.2.1 Non-realizable Representation with Two Transfer Functions

To account for the two different paths of $r(s)$ and $y(s)$ to $u(s)$ in (4.18), we can put forth a control law for ADRC using two transfer functions as follows, which is visualized in Fig. 4.2:

$$u(s) = C_{\text{FB}}(s) \cdot \left[C_{\text{PF}}^{\text{NR}}(s) \cdot r(s) - y(s) \right]. \tag{4.19}$$

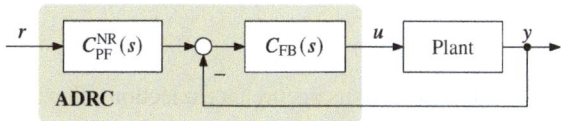

Fig. 4.2 Transfer function representation of a control loop with continuous-time ADRC, consisting of a feedback controller $C_{\text{FB}}(s)$ and a non-realizable prefilter $C_{\text{PF}}^{\text{NR}}(s)$

In (4.19), $C_{\text{FB}}(s)$ denotes a feedback controller transfer function, which can directly be read off (4.18):

$$C_{\text{FB}}(s) = \frac{\left(\boldsymbol{k}^{\text{T}}\ 1 \right) \cdot \operatorname{adj}(s\boldsymbol{I} - \boldsymbol{A}_{\text{CL}}) \cdot \boldsymbol{l}}{b_0 \cdot \det(s\boldsymbol{I} - \boldsymbol{A}_{\text{CL}})}. \tag{4.20}$$

The term $C_{\text{FB}}(s)$ is a strictly proper transfer function depending on the resolvent of $\boldsymbol{A}_{\text{CL}}$. In (4.16), it was made clear that $\boldsymbol{A}_{\text{CL}}$ has rank N, as its rightmost column is zero—assuming that both the controller and the observer gains are nonzero, of course. This implicates that the denominator polynomial $\det(s\boldsymbol{I} - \boldsymbol{A}_{\text{CL}})$ brings an integrator pole to $C_{\text{FB}}(s)$.

The second transfer function that we can extract from (4.18) describes a prefilter acting on the reference signal input $r(s)$:

$$C_{\text{PF}}^{\text{NR}}(s) = \frac{k_1 \cdot \det(s\boldsymbol{I} - \boldsymbol{A}_{\text{CL}}) - \frac{k_1}{b_0} \cdot \left(\boldsymbol{k}^{\text{T}}\ 1 \right) \cdot \operatorname{adj}(s\boldsymbol{I} - \boldsymbol{A}_{\text{CL}}) \cdot \boldsymbol{b}}{\left(\boldsymbol{k}^{\text{T}}\ 1 \right) \cdot \operatorname{adj}(s\boldsymbol{I} - \boldsymbol{A}_{\text{CL}}) \cdot \boldsymbol{l}}. \tag{4.21}$$

The order of the numerator in (4.21) is greater than the order of its denominator. Being improper—or not realizable—is the reason we attached a superscript "NR" to $C_{\text{PF}}^{\text{NR}}(s)$ in (4.19), (4.21), and Fig. 4.2.

The two transfer functions in (4.19), $C_{\text{FB}}(s)$ and $C_{\text{PF}}^{\text{NR}}(s)$, can be given in the following form, and we can clearly see that the denominator of $C_{\text{PF}}^{\text{NR}}(s)$ will cancel the zeros of $C_{\text{FB}}(s)$:

$$C_{\text{FB}}(s) = \frac{K_{\text{I}}}{s} \cdot \frac{1 + \sum_{i=1}^{N} \beta_i \cdot s^i}{1 + \sum_{i=1}^{N} \alpha_i \cdot s^i} \quad \text{and} \quad C_{\text{PF}}^{\text{NR}}(s) = \frac{1 + \sum_{i=1}^{N+1} \gamma_i \cdot s^i}{1 + \sum_{i=1}^{N} \beta_i \cdot s^i}. \tag{4.22}$$

For the cases most relevant to practice ($N = 1$ and $N = 2$), all α, β, γ coefficients mentioned in (4.22) and K_{I} are given in full detail in Appendix A.2.

Discussion

The core loop of LADRC, which is responsible for its disturbance rejection behavior, is formed by a feedback controller $C_{\text{FB}}(s)$, consisting of an integrator in series with an Nth-order lead-lag filter. For orders $N = 1$ and $N = 2$, $C_{\text{FB}}(s)$ is structurally equivalent to the so-called Type 2 or Type 3 controllers used in power electronics. These are equivalent to PI or PID controllers equipped with an additional first- or second-order low-pass filter, which is, in turn, a recommended practice in "classical" control engineering. A more detailed frequency-domain analysis of $C_{\text{FB}}(s)$ will be performed in Sect. 5.3.3 of this book.

While this structure of ADRC's transfer function representation is well known from classical control engineering—a feedback controller combined with a prefilter to shape the loop response—a downside remains with $C_{\text{PF}}^{\text{NR}}(s)$ being a non-realizable transfer function. This representation is therefore mainly suitable for mathematical analysis of a control loop. For this reason, we do not give a "cooking recipe" here—this form of ADRC cannot be "prepared" in the real world. Fortunately, a solution to that problem can be found and will be presented next.

4.2.2 A Realizable Transfer Function Representation

It is obvious from (4.22) that the problem consists of the s^{N+1} term in the numerator of $C_{\text{PF}}^{\text{NR}}(s)$. The prefilter would, however, be realizable if the order of its numerator polynomial could be reduced from $N+1$ to N. Let us assume that this is possible, and we could represent the prefilter with the following transfer function, which differs from (4.22) only in the omission of the $\gamma_{N+1} \cdot s^{N+1}$ term:

$$C_{\text{PF}}(s) = \frac{1 + \sum_{i=1}^{N} \gamma_i \cdot s^i}{1 + \sum_{i=1}^{N} \beta_i \cdot s^i}. \tag{4.23}$$

4.2 Transfer Function Representation

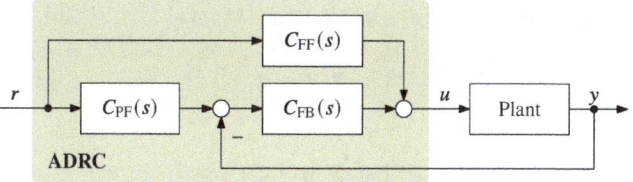

Fig. 4.3 Representation of control loop with continuous-time ADRC using realizable transfer functions, consisting of feedback controller $C_{FB}(s)$, prefilter $C_{PF}(s)$, and feedforward term $C_{FF}(s)$

Exactly that is possible. The non-realizable part of $C_{PF}^{NR}(s)$ can be implemented using a third transfer function bypassing the feedback controller $C_{FB}(s)$, as shown in Fig. 4.3—we split the realization of $C_{PF}^{NR}(s)$ up into two transfer functions $C_{PF}(s)$ and $C_{FF}(s)$. This new feedforward transfer function $C_{FF}(s)$ is now chosen to achieve the same transfer behavior from $r(s)$ to $u(s)$:

$$C_{PF}^{NR}(s) \cdot C_{FB}(s) \stackrel{!}{=} C_{PF}(s) \cdot C_{FB}(s) + C_{FF}(s).$$

Using (4.23) and (4.22), this gives

$$C_{FF}(s) = \left(C_{PF}^{NR}(s) - C_{PF}(s)\right) \cdot C_{FB}(s) = \frac{\gamma_{N+1} \cdot s^{N+1}}{1 + \sum_{i=1}^{N} \beta_i \cdot s^i} \cdot \frac{K_I}{s} \cdot \frac{1 + \sum_{i=1}^{N} \beta_i \cdot s^i}{1 + \sum_{i=1}^{N} \alpha_i \cdot s^i}.$$

After canceling matching terms, we obtain $C_{FF}(s)$ in form of a realizable Nth-order high-pass filter:

$$C_{FF}(s) = \frac{K_I \cdot \gamma_{N+1} \cdot s^N}{1 + \sum_{i=1}^{N} \alpha_i \cdot s^i}. \quad (4.24)$$

The two transfer functions $C_{FF}(s)$ and $C_{PF}(s)$ can also be derived from (4.18) with the same goal, yielding

$$C_{PF}(s) = \frac{k_1 \cdot \left(\det(s\boldsymbol{I} - \boldsymbol{A}_{CL}) - s^{N+1} - \frac{1}{b_0} \cdot (\boldsymbol{k}^T\ 1) \cdot \text{adj}(s\boldsymbol{I} - \boldsymbol{A}_{CL}) \cdot \boldsymbol{b}\right)}{(\boldsymbol{k}^T\ 1) \cdot \text{adj}(s\boldsymbol{I} - \boldsymbol{A}_{CL}) \cdot \boldsymbol{l}}$$

$$\text{and} \quad C_{FF}(s) = \frac{k_1 \cdot s^{N+1}}{b_0 \cdot \det(s\boldsymbol{I} - \boldsymbol{A}_{CL})}.$$

In summary, we can conclude that a realizable transfer function representation of continuous-time ADRC can be given with the following control law:

$$u(s) = C_{\text{FB}}(s) \cdot [C_{\text{PF}}(s) \cdot r(s) - y(s)] + C_{\text{FF}}(s) \cdot r(s) \qquad (4.25)$$

$$\text{with} \quad C_{\text{FB}}(s) = \frac{K_{\text{I}}}{s} \cdot \frac{1 + \sum_{i=1}^{N} \beta_i \cdot s^i}{1 + \sum_{i=1}^{N} \alpha_i \cdot s^i},$$

$$C_{\text{PF}}(s) = \frac{1 + \sum_{i=1}^{N} \gamma_i \cdot s^i}{1 + \sum_{i=1}^{N} \beta_i \cdot s^i}, \quad \text{and} \quad C_{\text{FF}}(s) = \frac{K_{\text{I}} \cdot \gamma_{N+1} \cdot s^N}{1 + \sum_{i=1}^{N} \alpha_i \cdot s^i}.$$

It is important to note that the same α, β, γ coefficients and K_{I} used in (4.22) of Sect. 4.2.1 are reused here for $C_{\text{FB}}(s)$, $C_{\text{PF}}(s)$, and $C_{\text{FF}}(s)$—they are only assigned to three instead of two transfer functions. Equations for obtaining these coefficients from an existing state-space tuning or from *bandwidth parameterization* are given in full detail in Appendix A.3.

Cooking Recipe (ADRC, Transfer Function Form)

To implement and tune the realizable transfer function form of ADRC derived in this section, the following steps have to be carried out:

1. *Plant modeling:* Identify the order N and the gain parameter b_0 of the plant model.
2. *Controller and observer tuning:* When using *bandwidth parameterization*, the tuning parameters are ω_{CL} and k_{ESO}. For first- and second-order ADRCs, these values can simply be put in the design equations for the K_{I} and α, β, γ coefficients tabulated in Table A.2.2 and Table A.2.3 on page 194, respectively. It is also possible to compute these coefficients from the otherwise obtained state-space controller and observer gains k^{T} and l.
3. *Implementation:* The structure of the control law (4.25) is shown in Fig. 4.3, consisting of a feedback controller $C_{\text{FB}}(s)$, a prefilter $C_{\text{PF}}(s)$, and a feedforward term $C_{\text{FF}}(s)$. The feedback controller $C_{\text{FB}}(s)$ contains an integrator, which, to avoid windup issues, should be implemented with output limits.

All relevant equations and details are once more compiled in the Appendix, for the non-realizable and this realizable transfer function representations in Appendices A.2 and A.3, respectively.

Open Access This chapter is licensed under the terms of the Creative Commons Attribution 4.0 International License (http://creativecommons.org/licenses/by/4.0/), which permits use, sharing, adaptation, distribution and reproduction in any medium or format, as long as you give appropriate credit to the original author(s) and the source, provide a link to the Creative Commons license and indicate if changes were made.

The images or other third party material in this chapter are included in the chapter's Creative Commons license, unless indicated otherwise in a credit line to the material. If material is not included in the chapter's Creative Commons license and your intended use is not permitted by statutory regulation or exceeds the permitted use, you will need to obtain permission directly from the copyright holder.

Chapter 5
Visual Tour

Abstract A picture is worth a thousand words. This chapter aims at analyzing and showcasing ADRC in a predominantly visual manner, in both time and frequency domains, going from modeling to tuning and its closed-loop behavior. What can one expect from ADRC regarding tracking performance and disturbance rejection? What is the influence of its modeling and tuning parameters? And what if we set them too low or too high?

5.1 Plant Modeling

5.1.1 Obtaining b_0 in Time Domain

We started introducing the core concepts of ADRC in Sect. 3.1 with its simple plant model (3.2) that—apart from the plant order N—only has one parameter b_0, the *critical gain parameter*. In this section, we want to demonstrate how b_0 can experimentally be obtained given a step response of a plant following a stepwise change in its input signal $u(t) = u_\infty \cdot \sigma(t)$, where $\sigma(t)$ is the Heaviside step function and u_∞ the steady-state value of the input—if applicable, relative to its initial value. This will be shown using three simple examples:

- *Plant with pure integrator behavior.* If the plant can be described as an integrator with a gain value K_I,

$$P(s) = \frac{K_I}{s} = \frac{b_0}{s},$$

its reaction to an input signal $u(t) = u_\infty \cdot \sigma(t)$ will be

$$y(t) = K_I \cdot t \cdot u_\infty \cdot \sigma(t) = b_0 \cdot t \cdot u_\infty \cdot \sigma(t).$$

As shown in case (a) of Fig. 5.1, b_0 can be computed from the normalized slope:

© The Author(s) 2025
G. Herbst, R. Madonski, *Active Disturbance Rejection Control*, Control Engineering, https://doi.org/10.1007/978-3-031-72687-3_5

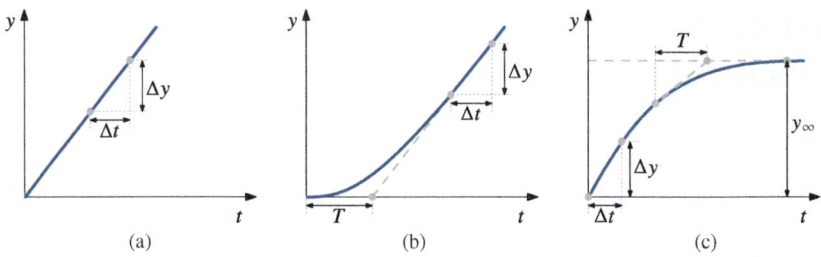

Fig. 5.1 Step responses of three simple, commonly seen plant types: (a) integrator, (b) integrator plus first-order lag, and (c) first-order low-pass. Using the specified characteristics, the *critical gain parameter* b_0 can be computed from such a visual plant description that might be obtained from experiments or simulations

$$b_0 = \frac{\Delta y}{\Delta t \cdot u_\infty}. \tag{5.1}$$

- *Plant with integrator plus first-order lag behavior.* If a plant has integrator behavior with a first-order lag (time constant T)

$$P(s) = \frac{K_I}{s} \cdot \frac{1}{Ts+1} = \frac{\frac{K_I}{T}}{s^2 + \frac{1}{T}s} = \frac{b_0}{s^2 + a_1 s},$$

the reaction to an input signal $u(t) = u_\infty \cdot \sigma(t)$ will be

$$y(t) = K_I \cdot \left(t - T \cdot \left(1 - e^{-\frac{t}{T}}\right)\right) \cdot u_\infty \cdot \sigma(t),$$

which, for $t \gg T$, approaches a ramp:

$$y(t \gg T) \approx K_I \cdot (t - T) \cdot u_\infty.$$

As shown in case (b) of Fig. 5.1, b_0 can then be computed from the normalized slope and the value of T read off the intersection of the ramp with the time axis:

$$b_0 = \frac{K_I}{T} = \frac{\Delta y}{\Delta t \cdot u_\infty \cdot T}. \tag{5.2}$$

- *Plant with first-order low-pass behavior.* For a self-regulating plant with first-order lag behavior (gain K and time constant T)

$$P(s) = \frac{K}{Ts+1} = \frac{\frac{K}{T}}{s + \frac{1}{T}} = \frac{b_0}{s + a_0},$$

the reaction to an input signal $u(t) = u_\infty \cdot \sigma(t)$ will be

$$y(t) = K \cdot \left(1 - e^{-\frac{t}{T}}\right) \cdot u_\infty \cdot \sigma(t).$$

There are two options to compute b_0. The first one uses K obtained from the steady-state value, $K = \frac{y_\infty}{u_\infty}$, and T from a tangent to $y(t)$ as depicted in Fig. 5.1, case (c). The value of b_0 then simply results from

$$b_0 = \frac{K}{T}. \tag{5.3}$$

The second option is to evaluate the step response at its onset, i.e., for $t \ll T$. At its beginning, $y(t)$ can be approximated by a linear response,

$$y(t) \approx \frac{K}{T} \cdot t \cdot u_\infty = b_0 \cdot t \cdot u_\infty, \quad 0 \leq t \ll T;$$

hence b_0 can be computed from the normalized slope:

$$b_0 = \frac{\Delta y}{\Delta t \cdot u_\infty}. \tag{5.4}$$

This can be extended to more complex plant dynamics, notably with second-order low-pass behavior, for which numerous identification methods exist to experimentally obtain their parameters. For ADRC, these can then be evaluated to compute b_0.

5.1.2 Understanding b_0 in Frequency Domain

Let us now turn to the frequency domain for a different interpretation of b_0 and another option to extract it from experimentally obtained data. The plant model (3.1) can, without the disturbance input, be given as a frequency-domain transfer function:

$$P(j\omega) = \frac{y(j\omega)}{u(j\omega)} = \frac{b_0}{(j\omega)^N + a_{N-1} \cdot (j\omega)^{N-1} + \ldots + a_1 \cdot j\omega + a_0}. \tag{5.5}$$

For higher frequencies, the denominator polynomial becomes dominated by its highest order term, and one can approximate

$$\lim_{\omega \to \infty} |P(j\omega)| \approx \left| \frac{b_0}{(j\omega)^N} \right| = \frac{b_0}{\omega^N}.$$

Solving this high-frequency approximation for its crossover frequency ω_X leads us to a relation between ω_X and b_0:

$$|P(j\omega_X)| = 0\,\text{dB} = 1 \quad \text{yields} \quad b_0 = \omega_X^N \quad \text{or} \quad \omega_X = \sqrt[N]{b_0}. \tag{5.6}$$

If an Nth-order plant can be modeled by (5.5), b_0 can therefore be found from a Bode magnitude plot of $|P(j\omega)|$ by approximating the upper-frequency range with an Nth-order integrator, i.e., a straight line. The 0 dB crossover frequency ω_X of this approximation then gives b_0 via (5.6). Various examples of first- and second-order plants with the same approximation to obtain b_0 are shown in Fig. 5.2.

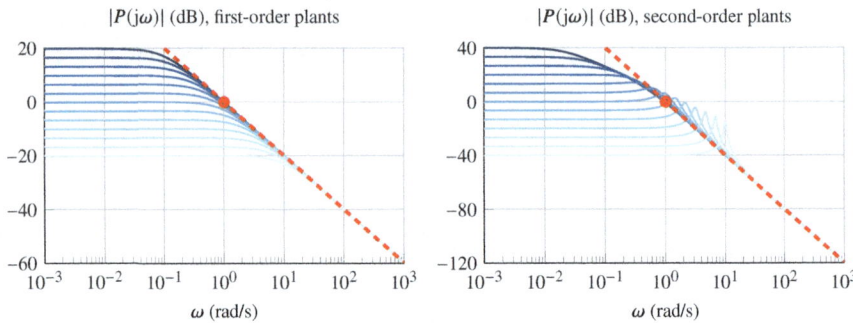

Fig. 5.2 The critical gain parameter b_0 in frequency domain: A variety of first- (left-hand side) and second-order plants (right-hand side plot) with the same value of $b_0 = 1$. First-order plants: $P(s) = \frac{b_0}{s+a_0}$ with a_0 ranging from 0.1 (darker colors) to 10 (brighter colors). Second-order plants: $P(s) = \frac{b_0}{s^2+0.5s+a_0}$ with a_0 ranging from 0.01 (darker colors) to 100 (brighter colors). b_0 can be obtained from a Bode magnitude plot $|P(j\omega)|$ by extending the straight-line approximation from higher frequencies (red dashed line) to find the 0 dB crossover frequency ω_X encircled in the diagrams and then computing $b_0 = \omega_X^N$ for an Nth-order plant

5.2 Closing the Loop: A Nominal Example

In a simple simulation example, which will serve as the nominal case of further examples studied in this chapter, a normalized, critically damped second-order plant with a transfer function $P(s) = \frac{1}{s^2+2s+1}$ shall be controlled. To do so, let us make use of the knowledge gained so far in the book and follow the "cooking recipe" for ADRC given on page 42:

1. *Plant modeling:* From $P(s)$ we obtain the plant order $N = 2$, which means we need a second-order ADRC. Recalling the examples in Table 3.1, we also determine the *critical gain parameter* as $b_0 = 1$.
2. *Controller tuning:* Since we are not given certain design goals or restrictions, we arbitrarily select a bandwidth of 0.2 Hz, i.e., $\omega_{CL} = 2\pi \cdot 0.2$ rad/s for this example. According to (3.21), this should result in a closed-loop settling time $T_{\text{settle},98\%} \approx 5$ s. The controller gains $k_{1/2}$ are then computed using the general design equation for *bandwidth parameterization* given by (3.19) or through consulting Table A.1.1.
3. *Observer tuning:* As a reasonable starting point, we select a relative observer bandwidth factor $k_{ESO} = 5$. To compute the observer gains $l_{1/2/3}$, we put ω_{CL} and k_{ESO} into the general design equation (3.24) or once more refer to Table A.1.1.
4. *Implementation:* In our model-based implementation, we use the state-space form of linear ADRC depicted in Fig. 3.5. We refer to this form as "standard" ADRC, whenever contrasting it with other or modified forms. It is the origin of all variants and implementations covered in this book.

Let us conduct a first test run. Note that the control loop for our simulation example, shown in Fig. 5.3, now includes additional inputs to model the effect of disturbances at the plant input (usually "load disturbance," d) and plant output

5.2 Closing the Loop: A Nominal Example

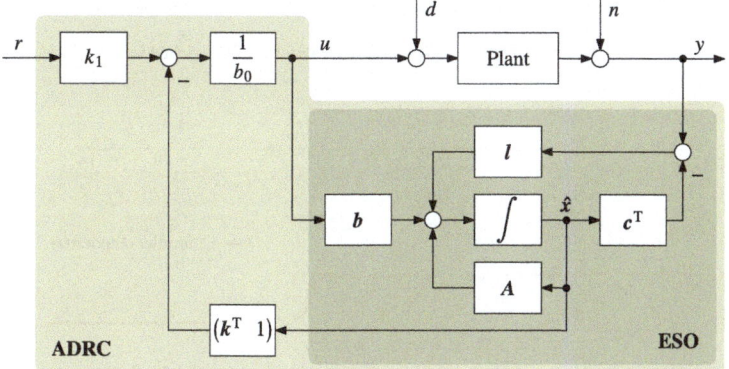

Fig. 5.3 Closed control loop with continuous-time ADRC, including disturbances at plant input (d, load disturbance) and plant output (n, measurement noise)

(measurement noise, n). At time $t = 0$, a stepwise change of the reference signal $r(t)$ from $r = 0$ to $r = 1$ occurs, followed by a stepwise change of the input disturbance $d(t)$ from $d = 0$ to $d = 0.5$ at $t = 10$ s.[1]

From the results shown in Fig. 5.4, one can nicely observe the closed-loop tracking behavior of $y(t)$ exhibiting the desired settling time. After $t = 10$ s, the controller output includes the efforts required to compensate for the persisting input disturbance. Remember that the control law (3.15) is $u(t) = \frac{1}{b_0} \cdot (k_1 \cdot r(t) - (k^T\ 1) \cdot \hat{x}(t))$. It is therefore very interesting to monitor the observer state variables $\hat{x}^T(t) = (\hat{x}_1(t)\ \hat{x}_2(t)\ \hat{x}_3(t))$, with $\hat{x}_3(t)$ being the estimated *total disturbance* $\hat{f}(t)$:

- *First phase*, $0\,\text{s} \leq t \lessapprox 5\,\text{s}$: During the transient following the reference step, all estimated states contribute to the controller output $u(t)$.
- *Second phase*, $5\,\text{s} \lessapprox t < 10\,\text{s}$: In the settled state, only the estimated *total disturbance* contributes to $u(t)$, as $r = y = \hat{x}_1$ and $\hat{x}_2 = 0$. The state variable $\hat{x}_3$ acts as the integrator state of this controller.
- *Third phase*, $10\,\text{s} \leq t$: When the input disturbance $d = 0.5$ becomes active at $t = 10$ s, $\hat{x}_3(t)$ changes to include the estimated input disturbance, such that $u(t)$ can compensate $d(t)$.

We have now successfully applied a "cooking recipe" to implement and tune ADRC and gained insights into the role of the observer state variables during reference tracking and disturbance rejection. In the remainder of this chapter, we will build on this example several times to demonstrate the influence of ADRC's tuning parameters and uncertainties to be expected in a not-so-well-known plant.

Please note that, in this nominal example, the closed loop is (approximately) only as slow or fast as the open-loop plant. This is a conscious decision on our side, as

---

[1] In this example, we will leave the second disturbance input $n(t) = 0$, as dealing with measurement noise is covered in a dedicated section later in this book—to be specific in Sect. 9.3, after discrete-time ADRC implementations have been introduced.

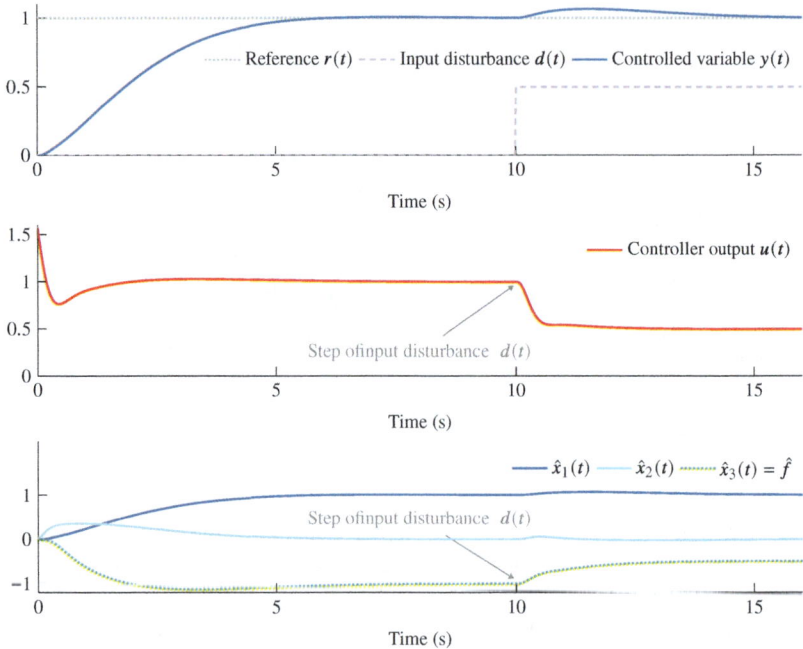

Fig. 5.4 Simulation example of the control loop from Fig. 5.3, showing the reaction of controlled variable $y(t)$, controller output $u(t)$, and observer state variables $\hat{x}_{1/2/3}(t)$ to a reference step and input disturbance step

we will need some headroom to compare this design to faster ones and with other varying parameters.

5.3 The Role of ADRC's Tuning Parameters

5.3.1 Preliminaries: The Gang of Six

Our focus in this chapter is to provide visual answers to "What happens (to the controller/control loop) if ..." questions, in order to create some intuition for the tuning parameters, the abilities, and the limitations of ADRC.

A block diagram of the control loop from Fig. 5.3, now in its realizable transfer function representation introduced in Sect. 4.2.2, is provided in Fig. 5.5. When answering the "What happens if ..." questions, the influence of the three possible input signals (reference r, load disturbance d, and measurement noise n) in Fig. 5.5 on both the controlled variable y and the controller output u will be of interest. We can give a compact representation of all possible transfer functions as follows, which are known as the *gang of six*:

5.3 The Role of ADRC's Tuning Parameters

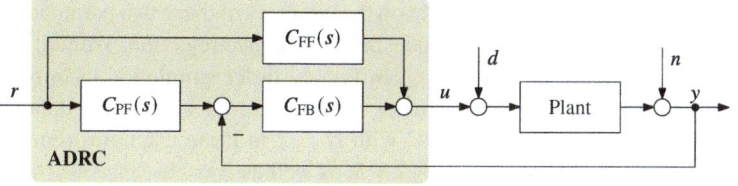

Fig. 5.5 Transfer function representation of a control loop with continuous-time ADRC, including disturbances at plant input (d) and plant output (n)

$$\begin{pmatrix} y(s) \\ u(s) \end{pmatrix} = \begin{pmatrix} G_{yr}(s) & G_{yd}(s) & G_{yn}(s) \\ G_{ur}(s) & G_{ud}(s) & G_{un}(s) \end{pmatrix} \cdot \begin{pmatrix} r(s) \\ d(s) \\ n(s) \end{pmatrix}. \tag{5.7}$$

The *gang of six* transfer functions can be computed from any transfer function representation of ADRC. Table 5.1 gives the *gang of six* in general form if the realizable representation from Sect. 4.2.2 is to be used. The particular transfer functions $C_{FB}(s)$, $C_{PF}(s)$, and $C_{FF}(s)$ have then to be inserted. In the remainder of this chapter, we will make use of the *gang of six*, showing some or all of these transfer functions in frequency domain (mostly as Bode magnitude plots) or time domain in the form of step responses.

Table 5.1 General form of the *gang of six* transfer functions for a control loop with a plant $P(s)$ and continuous-time ADRC using the realizable transfer function implementation consisting of feedback controller $C_{FB}(s)$, reference signal prefilter $C_{PF}(s)$, and feedforward $C_{FF}(s)$. Note that we omitted "(s)" from these transfer functions in the table only for lack of space

	From: r (reference signal)	From: d (disturbance)	From: n (noise)
To: y	$G_{yr}(s) = \dfrac{P \cdot (C_{FF} + C_{FB}C_{PF})}{1 + PC_{FB}}$	$G_{yd}(s) = \dfrac{P}{1 + PC_{FB}}$	$G_{yn}(s) = \dfrac{1}{1 + PC_{FB}}$
To: u	$G_{ur}(s) = \dfrac{C_{FF} + C_{FB}C_{PF}}{1 + PC_{FB}}$	$G_{ud}(s) = \dfrac{-PC_{FB}}{1 + PC_{FB}}$	$G_{un}(s) = \dfrac{-C_{FB}}{1 + PC_{FB}}$

5.3.2 On Bandwidth: Reference Tracking Versus Disturbance Rejection

As described in Sects. 3.2.1 and 3.2.2, the process of tuning ADRC involves parameterization of an outer state-feedback controller and of an observer at its core, the latter providing estimated states for both the state-feedback controller and disturbance rejection. Using the *bandwidth parameterization* approach, this task is reduced to choosing a closed-loop bandwidth ω_{CL} and a relative bandwidth factor k_{ESO} for the observer.

In this section, we want to shed some light on the interplay of these two parameters and their impact on the tracking performance and disturbance rejection. We will do so with parameter sweeps of ω_{CL} and k_{ESO} in second-order simulation examples for the normalized plant $P(s) = \frac{1}{s^2+2s+1}$ already introduced in Sect. 5.2. The default tuning and modeling parameters for ADRC with $N = 2$ in these examples remain the same with $\omega_{CL} = 2\pi \cdot 0.2\,\text{rad/s}$ and $k_{ESO} = 5$, as well as $b_0 = 1$.

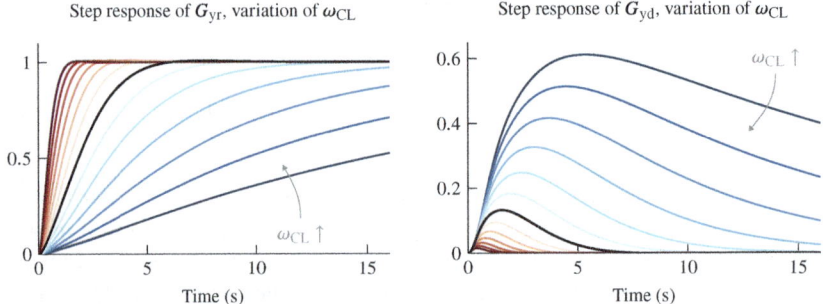

Fig. 5.6 Influence of ω_{CL} on closed-loop step responses of $G_{yr}(s)$ (tracking performance) and $G_{yd}(s)$ (disturbance rejection) for second-order ADRC with a normalized plant $P(s) = \frac{1}{s^2+2s+1}$. Values for ω_{CL} are logarithmically spaced, ranging from $\omega_{CL} = 2\pi \cdot 0.06\,\text{rad/s}$ (0.06 Hz, blue) through $\omega_{CL} = 2\pi \cdot 0.2\,\text{rad/s}$ (0.2 Hz, thick black line) to $\omega_{CL} = 2\pi \cdot 0.\bar{6}\,\text{rad/s}$ (0.6 Hz, red). Other tuning parameters remain fixed at $b_0 = 1$ and $k_{ESO} = 5$

A first set of simulation results is shown in Fig. 5.6: reactions to reference signal and input disturbance steps, i.e., step responses of $G_{yr}(s)$ and $G_{yd}(s)$ from the *gang of six*. Clearly, increasing the bandwidth ω_{CL} results in both improved tracking performance and better disturbance rejection at the same time.[2]

A second set of examples presented in Fig. 5.7 fixes ω_{CL} and varies k_{ESO}, i.e., only the observer speed. As visible from the step responses of $G_{yr}(s)$, the tracking performance suffers and does not meet the desired design goals (settling time) if k_{ESO} is chosen significantly below $k_{ESO} = 5$. On the other hand, increasing k_{ESO} beyond the single-digit range does not endlessly improve the tracking performance, as the controller can obviously not do more than perfectly achieving the design goal. However, the impact of the input disturbance $d(t)$ on the controlled variable $y(t)$ is obviously reduced increasingly better with larger values of k_{ESO}.

What can we learn from these examples?

- Tracking and rejection are *both* influenced by ω_{CL} *and* k_{ESO}. This means that ADRC, as introduced in Chap. 3, does not allow for a true two-degrees-of-freedom (2DOF) design with separate goals for tracking and disturbance rejection. This impression could, however, have arisen earlier when we introduced the transfer function representation in Sect. 4.2. Structurally, ADRC is a 2DOF controller, but

---

[2] Note that, since $k_{ESO} = 5$ is fixed in the examples of Fig. 5.7, the bandwidth of both controller and observer is increasing at the same relative pace.

5.3 The Role of ADRC's Tuning Parameters

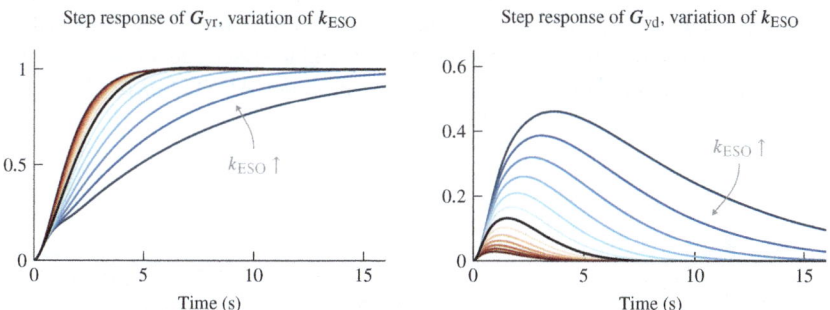

Fig. 5.7 Influence of k_{ESO} on closed-loop step responses of $G_{yr}(s)$ (tracking performance) and $G_{yd}(s)$ (disturbance rejection) for second-order ADRC with a normalized plant $P(s) = \frac{1}{s^2+2s+1}$. Values for k_{ESO} are logarithmically spaced, ranging from $k_{ESO} = 1$ (blue) through $k_{ESO} = 5$ (thick black line) to $k_{ESO} = 25$ (red). Other tuning parameters remain fixed at $b_0 = 1$ and $\omega_{CL} = 2\pi \cdot 0.2\,\text{rad/s}$ (i.e., $0.2\,\text{Hz}$)

the two design goals are intertwined in the tuning parameters. Note that this does not depend on a particular tuning approach, such as *bandwidth parameterization*. As can be seen in the coefficients of the transfer function representation in Appendix A.2, all transfer functions depend on both controller *and* observer gains. Modifying these gains therefore affects the performance of reference tracking and disturbance rejection at the same time.

- Is it possible yet to conclude on a best choice for ω_{CL} and k_{ESO}? Harder, better, faster, stronger? The obvious benefits of increased bandwidth must be paid for with increased feedback gains, making the controller and observer more susceptible to measurement noise and stability problems. In Sect. 5.3.3 we will make this compromise much more apparent when analyzing the influence of the observer bandwidth in greater detail.

5.3.3 Influence of the Observer Bandwidth

Gang-of-Six Analysis

In Sect. 3.1, we explicated the central role of the observer within ADRC. To somewhat decouple its tuning procedure from that of the controller, the tuning factor k_{ESO} was introduced within the *bandwidth parameterization* approach outlined in Sect. 3.2.2, determining the relative bandwidth of the observer compared to the closed control loop. Its influence on ADRC's performance will be examined in the following. This will also help answering the question raised in Sect. 5.3.2: What are the downsides of using a large relative observer bandwidth k_{ESO}, when the benefits of doing so appear so obvious?

To provide a full picture of k_{ESO}'s influence on the closed loop dynamics, a frequency-domain analysis of the *gang of six* transfer functions will be performed, with a parameter sweep of k_{ESO} for each of the transfer functions. This will be once more exemplified based on the nominal example described in Sect. 5.2, i.e., for a normalized plant $P(s) = \frac{1}{s^2+2s+1}$ and second-order ADRC tuned with $\omega_{CL} = 2\pi \cdot 0.2$ rad/s, $k_{ESO} = 5$, and $b_0 = 1$.

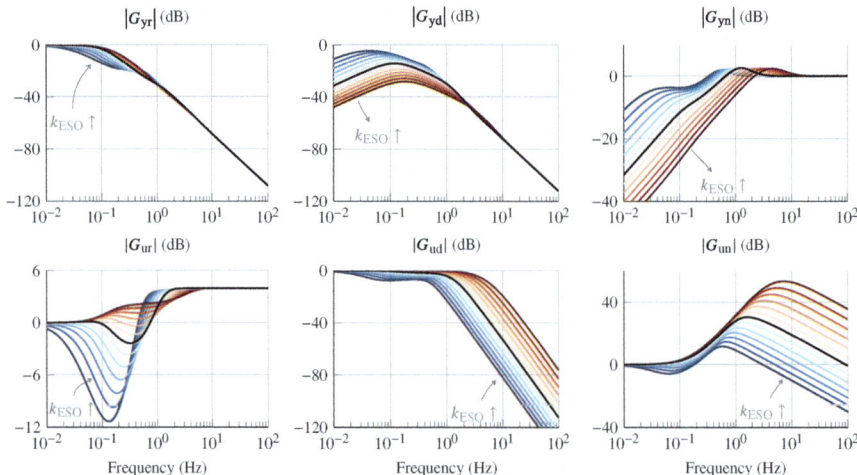

Fig. 5.8 Influence of k_{ESO} variations on the *gang of six* transfer functions for second-order ADRC. Normalized plant: $P(s) = \frac{1}{s^2+2s+1}$. Fixed tuning parameters are $b_0 = 1$ and $\omega_{CL} = 0.4\pi$ (i.e., 0.2 Hz). Values of k_{ESO} are logarithmically spaced, ranging from $k_{ESO} = 1$ (blue) through $k_{ESO} = 5$ (thick black line) to $k_{ESO} = 25$ (red)

The results (Bode magnitude plots) are presented in Fig. 5.8. Note that the step responses previously shown in Fig. 5.7 are the time-domain counterparts of $|G_{yr}(j\omega)|$ and $|G_{yd}(j\omega)|$ in Fig. 5.8, which confirm the positive impact of increasing k_{ESO} values on achieving the desired closed-loop bandwidth and on rejecting low-frequency disturbances. The biggest price to be paid for these benefits can be seen from $|G_{un}(j\omega)|$: With increasing k_{ESO}, the controller output u will be increasingly susceptible to high-frequency disturbances.

We have not yet elaborated on the influence and treatment of measurement noise—this topic will be covered in Sect. 9.3—but nonetheless hope all readers agree on our view by now that measurement noise will be inevitable in any practical setting. In the particular example of Fig. 5.8, increasing k_{ESO} from 5 to 25 (i.e., by a factor of 5) results in almost a 40 dB gain increase in $|G_{un}(j\omega)|$ for high frequencies. With $k_{ESO} = 25$, the noise level in the controller output would be larger by almost two orders of magnitude. In practical settings, this can be a hard restriction on the range of possible values for k_{ESO}, and this problem becomes worse with increasing ADRC order N.

The compromise one has to find between good tracking performance and low-frequency disturbance rejection on the one hand, and susceptibility to high-frequency

5.3 The Role of ADRC's Tuning Parameters

disturbances on the other, is, of course, nothing new: "Sensitivity reduction at low frequency unavoidably leads to sensitivity increase at higher frequencies."[3]

Frequency-Domain Analysis of the Feedback Controller

As visible from Fig. 5.5, the feedback controller $C_{FB}(s)$ in the transfer function representation of ADRC plays a central role in stability, disturbance rejection, and susceptibility to measurement noise. The "what if" question in this section therefore is: What happens to the feedback controller $C_{FB}(s)$ if one increases or decreases the tuning parameter k_{ESO}? The visual answer is given in the form of Bode plots in Fig. 5.9 for the first- and second-order cases.

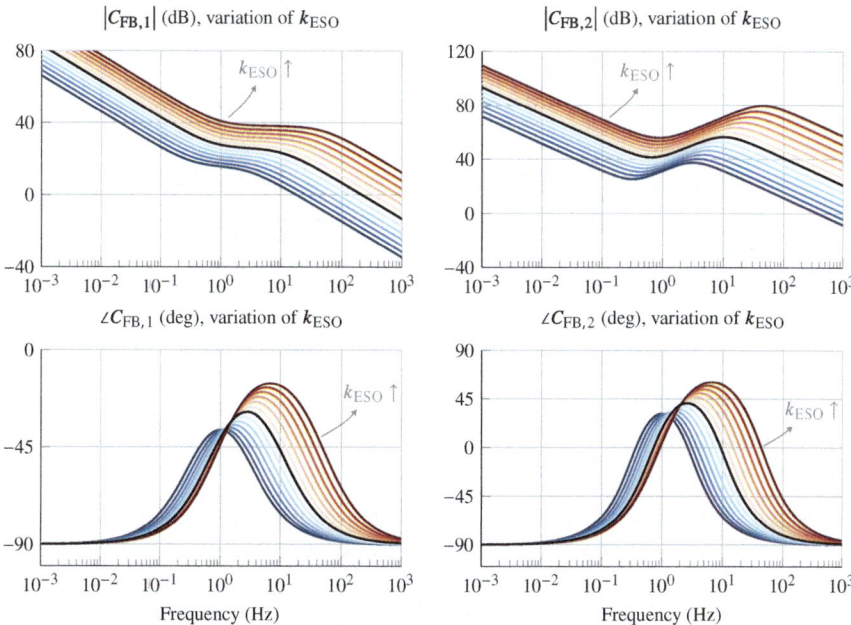

Fig. 5.9 Bode plots of the feedback controller $C_{FB}(s)$ for first- and second-order ADRCs. Values for k_{ESO} are logarithmically spaced, ranging from $k_{ESO} = 1$ (blue) through $k_{ESO} = 5$ (black) to $k_{ESO} = 25$ (red). Other tuning parameters remain fixed at $b_0 = 1$ and $\omega_{CL} = 2\pi$ (for a desired bandwidth of 1 Hz)

To interpret these Bode plots, recall the transfer functions for the first- and second-order ADRCs from (4.22) in Sect. 4.2:

$$C_{FB,1}(s) = \frac{K_I}{s} \cdot \frac{1 + \beta_1 s}{1 + \alpha_1 s} \quad \text{and} \quad C_{FB,2}(s) = \frac{K_I}{s} \cdot \frac{1 + \beta_1 s + \beta_2 s^2}{1 + \alpha_1 s + \alpha_2 s^2}.$$

---

[3] Cf. Gunter Stein's famous Bode Lecture "Respect the Unstable" at the 1989 IEEE Conference on Decision and Control.

Obviously the $C_{FB}(s)$ transfer functions consists of an integrator $\frac{K_I}{s}$ in series with Nth-order lead-lag filters. From Fig. 5.9, one can recognize that the latter only act as lead filters, increasingly providing phase boost (not exactly) around ω_{CL} for larger values of k_{ESO}. These feedback controllers are the "Type 2" and "Type 3" controllers, already mentioned in Sect. 4.2.1, which are, in turn, only special cases of PI and PID controllers with an additional first- or second-order low-pass filter, respectively. Note that the existence of this low-pass component can be attributed to the choice of using a full-order observer in the variants of ADRC considered in this book: The outer controller feeds back the estimated value $\hat{y}$ of the plant output y, even though the latter is measured and could be used directly, leading to a reduced-order observer. Estimating $\hat{y}$, on the other hand, comes with low-pass filter characteristics, which are beneficial for reducing the impact of measurement noise.

Let us now go into some more detail for first-order ADRC by putting the coefficients for *bandwidth parameterization* from Table A.2.2 in $C_{FB,1}(s)$:

$$C_{FB,1}(s) = \underbrace{\frac{1}{b_0} \cdot \frac{k_{ESO}^2 \omega_{CL}^2}{1 + 2k_{ESO}}}_{K_I} \cdot \frac{1}{s} \cdot \frac{1 + \overbrace{\frac{2 + k_{ESO}}{k_{ESO}\omega_{CL}}}^{\beta_1 = 1/\omega_Z} \cdot s}{1 + \underbrace{\frac{1}{\omega_{CL} \cdot (1 + 2k_{ESO})}}_{\alpha_1 = 1/\omega_P} \cdot s}. \qquad (5.8)$$

It may become increasingly clear why we focus on analyzing the influence of k_{ESO} on $C_{FB}(s)$. Changing b_0, as part of K_I, merely moves the magnitude plots up or down, while changing ω_{CL} moves the pole and zero frequencies ω_P and ω_Z to the left or the right. Both visually from Fig. 5.9 and analytically from (5.8), one can make the following observations regarding k_{ESO}:

- As part of K_I, increasing/decreasing k_{ESO} moves the magnitude plot up/down. For larger values of k_{ESO}, this approximately becomes a proportional influence:

$$\lim_{k_{ESO} \to \infty} K_I = \lim_{k_{ESO} \to \infty} \frac{1}{b_0} \cdot \frac{k_{ESO}^2 \omega_{CL}^2}{1 + 2k_{ESO}} = \frac{1}{b_0} \cdot \frac{k_{ESO} \omega_{CL}^2}{2}.$$

- With increasing values of k_{ESO}, the angular frequency ω_Z of the zero in $C_{FB,1}(s)$ converges to the closed-loop bandwidth ω_{CL}:

$$\lim_{k_{ESO} \to \infty} \omega_Z = \lim_{k_{ESO} \to \infty} \frac{k_{ESO} \omega_{CL}}{2 + k_{ESO}} = \omega_{CL}.$$

- With increasing values of k_{ESO}, the angular frequency ω_P of the pole in $C_{FB,1}(s)$, i.e., the cutoff frequency of its low-pass filter component, goes to infinity:

$$\lim_{k_{ESO} \to \infty} \omega_P = \lim_{k_{ESO} \to \infty} \omega_{CL} \cdot (1 + 2k_{ESO}) \to \infty.$$

5.3 The Role of ADRC's Tuning Parameters

Since the passband of the low-pass filter grows with k_{ESO}, the influence of high-frequency measurement noise on the controller output will increase—a behavior we could already spot in $|G_{un}|$ of Fig. 5.8.

We can summarize these observations by stating that, for $k_{ESO} \to \infty$, the feedback controller $C_{FB,1}(s)$ of LADRC would converge to a standard PI controller only consisting of an integrator and a zero at $-\omega_{CL}$. A similar analysis can be performed for the feedback controller component of second-order ADRC:

$$C_{FB,2}(s) = \underbrace{\frac{1}{b_0} \cdot \frac{k_{ESO}^3 \omega_{CL}^3}{1 + 6k_{ESO} + 3k_{ESO}^2} \cdot \frac{1}{s}}_{K_I}$$

$$\cdot \frac{1 + \overbrace{\frac{1}{\omega_{CL}} \cdot \left(\frac{3}{k_{ESO}} + 2\right)}^{\beta_1} \cdot s + \overbrace{\frac{1}{\omega_{CL}^2} \cdot \left(\frac{3}{k_{ESO}^2} + \frac{6}{k_{ESO}} + 1\right)}^{\beta_2} \cdot s^2}{1 + \underbrace{\frac{2 + 3k_{ESO}}{\omega_{CL} \cdot \left(1 + 6k_{ESO} + 3k_{ESO}^2\right)}}_{\alpha_1} \cdot s + \underbrace{\frac{1}{\omega_{CL}^2 \cdot \left(1 + 6k_{ESO} + 3k_{ESO}^2\right)}}_{\alpha_2} \cdot s^2}.$$

Similar to $C_{FB,1}(s)$, $C_{FB,2}(s)$ consists of an integrator and a lead-lag filter, this time of order 2. Solving the roots of the numerator and denominator polynomials of the latter gives a pair of conjugate complex poles $s_{Z1/2}$ (for the zeros) and $s_{P1/2}$ (for the poles):

$$s_{Z1/2} = \frac{k_{ESO}\omega_{CL}}{3 + 6k_{ESO} + k_{ESO}^2} \cdot \left[-\left(\frac{3}{2} + k_{ESO}\right) \pm j\sqrt{\frac{3}{4} + 3k_{ESO}}\right], \quad \text{and}$$

$$s_{P1/2} = \omega_{CL} \cdot \left[-\left(1 + \frac{3}{2}k_{ESO}\right) \pm j\sqrt{3k_{ESO} + \frac{3}{4}k_{ESO}^2}\right].$$

As for the first-order case, the influence of k_{ESO} on the integrator gain and the position of zeros and poles of $C_{FB,2}(s)$ can now be analyzed:

- Increasing or decreasing k_{ESO} moves the magnitude plot up or down. This influence converges to a linear one for large values of k_{ESO}:

$$\lim_{k_{ESO} \to \infty} K_I = \lim_{k_{ESO} \to \infty} \frac{1}{b_0} \cdot \frac{k_{ESO}^3 \omega_{CL}^3}{1 + 6k_{ESO} + 3k_{ESO}^2} = \frac{1}{b_0} \cdot \frac{k_{ESO}\omega_{CL}^3}{2}.$$

- With increasing values of k_{ESO}, the two zeros converge to a pair of real-valued zeros with a common angular frequency at the closed-loop bandwidth ω_{CL}:

$$\lim_{k_{ESO} \to \infty} s_{Z1/2} = -\omega_{CL}.$$

- The poles $s_{P1/2}$ remain conjugate complex for increasing values of k_{ESO}, with an approximately constant damping ratio that starts at $D \approx 0.791$ for $k_{ESO} = 1$ and relatively quickly converges to $D = \sqrt{3/4} \approx 0.866$.[4] However, at the same time, the real part of $s_{P1/2}$ moves toward minus infinity. This means that high-frequency measurement noise will increasingly affect the controller output.

Summarizing these findings we can therefore conclude that the feedback controller $C_{FB,2}(s)$ of LADRC converges to a standard PID controller with $k_{ESO} \to \infty$.

5.3.4 Influence of the Critical Gain Parameter

In Sect. 3.1, we have introduced b_0 as the *critical gain parameter*—which is, besides the plant order N, the only modeling information required for ADRC. We have presented several ways to obtain b_0 apart from a mathematical model in Sect. 5.1. But even if modeling efforts are largely reduced that way, what happens if one over- or underestimates b_0?

As in Sect. 5.3.3, we want to give a visual answer to this question by applying a parameter sweep of b_0 to the *gang of six* transfer functions. This will, again, be based on the nominal example from Sect. 5.2, i.e., for a normalized plant $P(s) = \frac{1}{s^2+2s+1}$ and second-order ADRC tuned with $\omega_{CL} = 2\pi \cdot 0.2\,\text{rad/s}$ and $k_{ESO} = 5$. The exact value of b_0 is therefore $b_0 = 1$.

Figure 5.10 contains the Bode magnitude plots of the *gang of six* and Fig. 5.11 selected time-domain step responses to complement their interpretation. We can summarize these results in two major takeaways:

- *Overestimating* b_0 will result in undermatching the desired bandwidth. This means that the tracking performance will suffer, and disturbances will be compensated more slowly.
- *Underestimating* b_0, on the other hand, will result in a more aggressively tuned controller. This could already be expected from the $\frac{1}{b_0}$ gain factor in the feedback controller transfer function $C_{FB}(s)$. In the *gang of six* magnitude plots, one can recognize the effects of this: increased sensitivity of the controller output u to measurement noise n, and an increasingly pronounced resonance peak emerging in all closed-loop transfer functions. One can particularly well observe this in the time-domain step responses of the controller output $u(t)$, where oscillations (in these examples roughly around 2 Hz) are increasingly dominating the response for significantly underestimated values of b_0. Ultimately, severe underestimation of b_0 will therefore result in instability.

---

[4] This is a slightly underdamped design falling into a range very typical for the damping D of a second-order noise filter ($D \in [1/\sqrt{2}, 1]$).

5.3 The Role of ADRC's Tuning Parameters

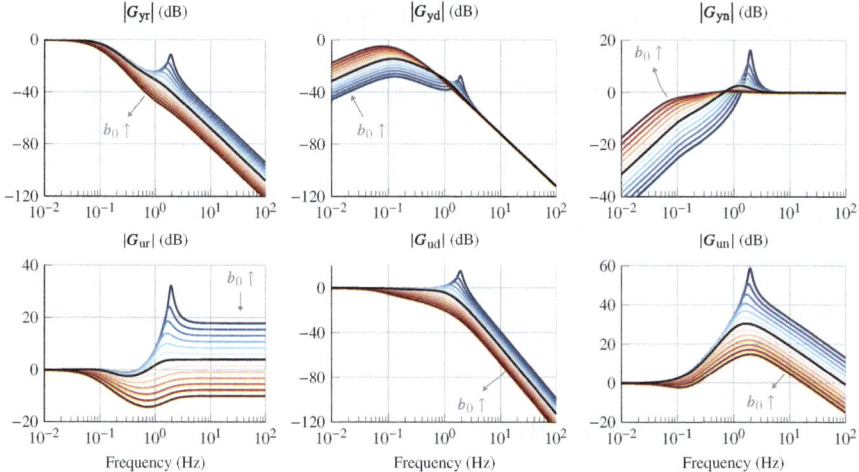

Fig. 5.10 Influence of under- or overestimating b_0 on the *gang of six* transfer functions, in this example for second-order ADRC. Normalized plant: $P(s) = \frac{1}{s^2+2s+1}$. Values of b_0 are logarithmically spaced, ranging from $b_0 = 0.2$ (blue, underestimation) through $b_0 = 1$ (thick black line, nominal value) to $b_0 = 5$ (red, overestimation). Other tuning parameters remain fixed at $\omega_{\text{CL}} = 2\pi \cdot 0.2 \,\text{rad/s}$ (i.e., 0.2 Hz) and $k_{\text{ESO}} = 5$

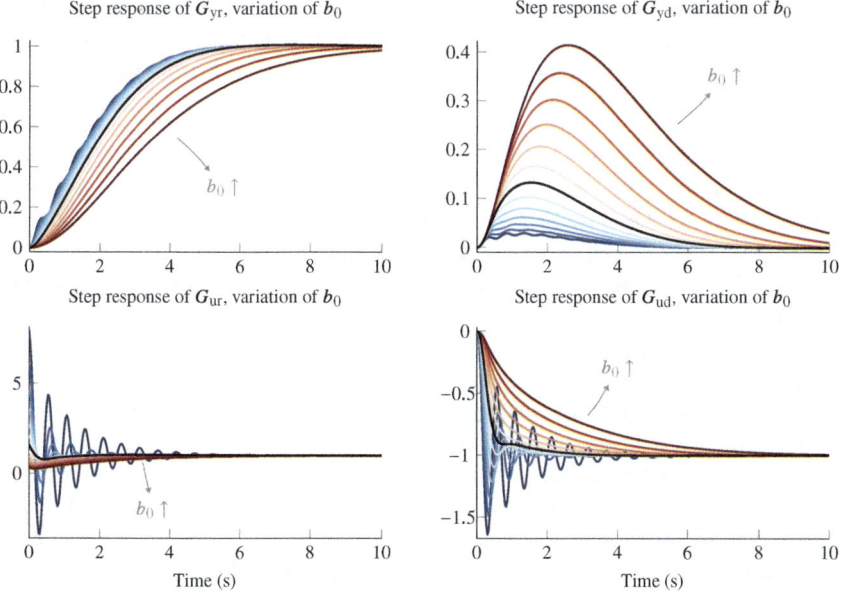

Fig. 5.11 Closed-loop responses to stepwise changes of the reference signal r and the input disturbance d, i.e., step responses of four of the six transfer functions from Fig. 5.10. Values of b_0 are logarithmically spaced, ranging from $b_0 = 0.2$ (blue, underestimation) through $b_0 = 1$ (thick black line, nominal value) to $b_0 = 5$ (red, overestimation)

5.4 Coping with Parametric and Structural Plant Uncertainties

In Sect. 5.3.4 we have already seen what happens if the estimated value of b_0 does not match the actual value of the plant. However, plant dynamics are not only dependent on this single, lumped parameter—others were just discarded in the process of ADRC's simplified plant modeling approach. Besides unexpected parameter changes, there may also be unmodeled dynamics in a plant—such as additional poles or zeros. To develop an intuition of how ADRC will handle such cases, we will perform a deeper analysis based on time-domain examples in two parts in the following.

5.4.1 Variation of Plant Parameters

One can view b_0 as a lumped estimation of some of a plant's parameters, cf. Table 3.1. What happens if individual plant parameters diverge from this estimation, regardless of if they are part of b_0?

Let us once more consider our normalized example from Sect. 5.2, a second-order plant $P(s) = \frac{1}{s^2+2s+1}$. As per Table 3.1, this plant has a gain $K = 1$, damping $D = 1$, and time constant $T = 1$ s. We will now perform parameter sweeps, changing each of the plant's parameters, and provide closed-loop step responses of $G_{yr}(s)$ and $G_{ur}(s)$—i.e., the controlled variable and the controller output. All simulations are performed with second-order ADRC tuned with fixed settings $\omega_{CL} = 2\pi \cdot 0.2$ rad/s, $k_{ESO} = 5$, and $b_0 = 1$.

- *Varying plant gain K:* As visible from Fig. 5.12, an increasing value of K has the same effect on the closed-loop step response of $y(t)$, as the underestimation of b_0 in Fig. 5.11. The effect on the controller output $u(t)$ is different, of course, as its steady-state value now must cater to the varying plant gain. When the plant gain K exceeds the value assumed in b_0, the control loop may ultimately become unstable. A reduced plant gain, on the other hand, will result in a closed-loop bandwidth that does not match the intended design.
- *Varying plant damping D:* It is quite interesting to see in Fig. 5.13 that the control loop is relatively insensitive to changes of the plant damping D, even though this value was varied by a factor of five in both directions in these examples. The slowed-down, increasingly overshooting response for large values of D results from the large time constant emerging in the denominator of the plant transfer function: $1 + 2DTs + T^2s^2 = (1+T_1s) \cdot (1+T_2s)$ for $D > 1$, with $T_{1/2} = \frac{T}{D \pm \sqrt{D^2-1}}$. For such heavily overdamped cases, it might be favorable to resort to first-order ADRC, focusing only on the larger time constant.
- *Varying plant time constant T:* As visible from Fig. 5.14, the effect of a varying plant time constant ranges from oscillations for reduced values of T (ultimately, instability) to slowed-down, overshooting responses for larger values of T. Note that the range of examples chosen for T variations was reduced to a factor of 3 in

5.4 Coping with Parametric and Structural Plant Uncertainties

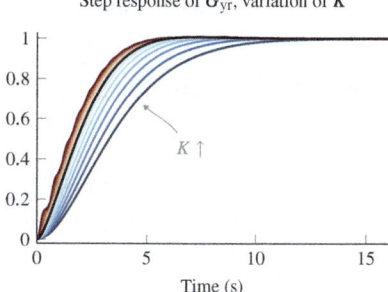
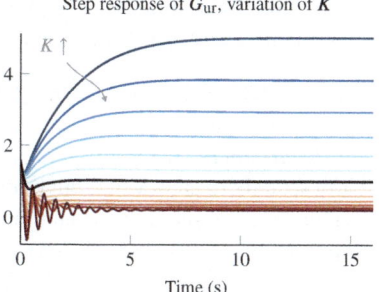

Fig. 5.12 Effect of varying gain K of a plant $P(s) = \frac{K}{s^2+2s+1}$ on the closed-loop step responses of controlled variable y and controller output u. Controller parameters remain fixed, designed for the nominal case $K = 1$. Values of K are logarithmically spaced, ranging from $K = 0.2$ (blue) through $K = 1$ (thick black line, nominal value) to $K = 5$ (red)

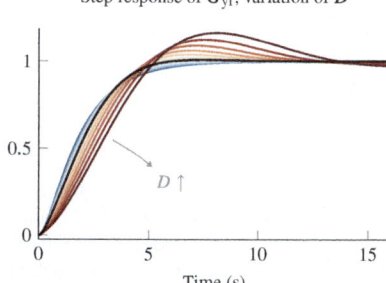
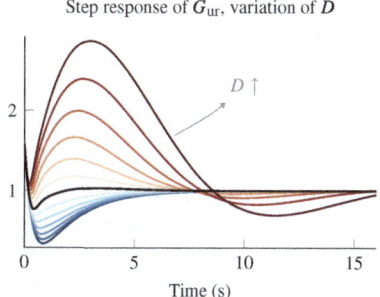

Fig. 5.13 Effect of varying gain D of a plant $P(s) = \frac{1}{s^2+2Ds+1}$ on the closed-loop step responses of controlled variable y and controller output u. Controller parameters remain fixed. Values of D are logarithmically spaced, ranging from $D = 0.2$ (blue) through $D = 1$ (thick black line, nominal value) to $D = 5$ (red)

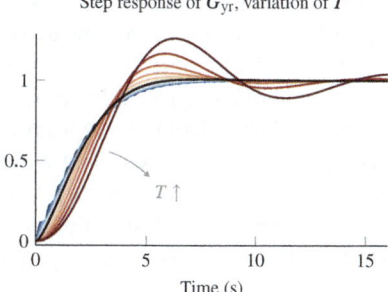
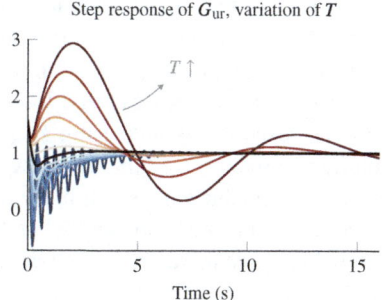

Fig. 5.14 Effect of varying gain T of a plant $P(s) = \frac{1}{T^2s^2+2Ts+1}$ on the closed-loop step responses of controlled variable y and controller output u. Controller parameters remain fixed, designed for the nominal case $T = 1$. Values of T are logarithmically spaced, ranging from $T = \frac{1}{3}$ s (blue) through $T = 1$ s (thick black line, nominal value) to $T = 3$ s (red)

both directions, since $b_0 = \frac{K}{T^2}$, i.e., the time constant has a quadratic influence on the plant's critical gain parameter.

In Sect. 3.1 we started motivating the ingredients of ADRC by stating that unmodeled plant dynamics or parameter variations will be compensated by the estimated *total disturbance*—if the observer is fast enough. In the examples above we have clearly seen there are (even though reasonable) limits to this ability, at least when using a moderate observer bandwidth factor $k_{ESO} = 5$. Let us therefore briefly consider an obvious fix to this problem: speeding up the observer.

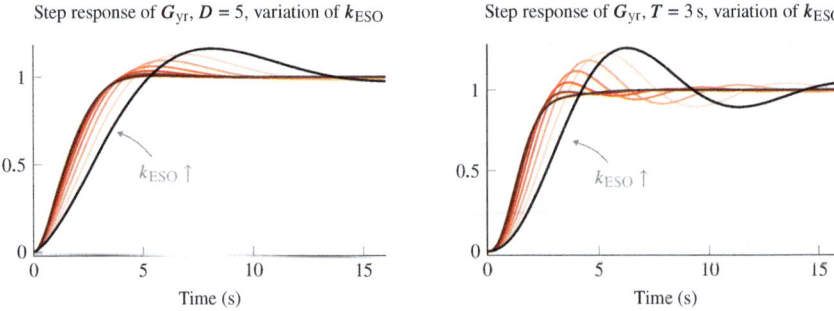

Fig. 5.15 High-gain observers: a quick fix to large plant uncertainties? Two examples start from worst-case deviations from the nominal plant $P(s) = \frac{1}{s^2+2s+1} = \frac{K}{T^2s^2+2DTs+1}$ in Fig. 5.13 (damping increased to $D = 5$) and Fig. 5.14 (time constant increased to $T = 3$ s). Increasing the relative observer bandwidth k_{ESO} seemingly fixes the problem, helping to maintain the desired tracking performance. In practice, however, k_{ESO} cannot usually be increased to such large bandwidths due to stability issues and the influence of measurement noise, as discussed in Sect. 5.3.3 and Sect. 9.3. Values of k_{ESO} are logarithmically spaced, beginning with the nominal case ($k_{ESO} = 5$, thick black line), ranging to $k_{ESO} = 50$ (red colors, increasingly dark)

In Fig. 5.15 the largest deviations of D and T from Figs. 5.13 and 5.14 are picked up again, this time with increasing values of k_{ESO}. Going beyond the range considered in Sect. 5.3.3, k_{ESO} is increased up to a value of $k_{ESO} = 50$ in these examples. As can be seen, the resulting closed-loop step responses are brought back to the desired nominal response. Is increasing k_{ESO} the silver bullet? While looking like a great solution in theory, we already know the answer from Sect. 5.3.3: Increasing k_{ESO} beyond reasonable values will make the controller very susceptible to measurement noise and even more so for higher orders N. The maximum allowable observer bandwidth must therefore be individually found for each application.

5.4.2 Additional Plant Pole or Zero

Beyond unknown or varying parameters of a plant to be controlled, there will often also be unmodeled dynamics—i.e., there are more poles or zeros in the actual plant

5.4 Coping with Parametric and Structural Plant Uncertainties

than considered when setting up the (coarse) model for ADRC. As an example, a process may exhibit a predominant second-order behavior, yet there are higher-order dynamics, such as a minor time constant introduced by a limited-bandwidth sensor.

What can one expect from the closed-loop behavior of ADRC when encountering such situations? We want to consider two scenarios, using—for the last time in this chapter— the nominal example introduced in Sect. 5.2.

- *Additional pole:* In the first scenario, an additional (unknown) pole with a time constant T_P is being added to the normalized plant from Sect. 5.2, i.e., plant $P(s)$ now reads

$$P(s) = \frac{1}{s^2 + 2s + 1} \cdot \frac{1}{T_P s + 1}.$$

As T_P is assumed to be unknown when designing the controller, the plant is only modeled with $N = 2$ and $b_0 = 1$. Figure 5.16 shows several examples with values of T_P ranging up to $T_P = 1$ s—quite a large range of values considered, as the plant will have three identical poles at in the latter case, yet is treated as a second-order plant. One can attest that this control loop is pretty insensitive to additional minor dynamics, supporting the initial claim of ADRC that a coarse model for the dominant dynamics indeed is a reasonable starting point in the process of designing the controller.

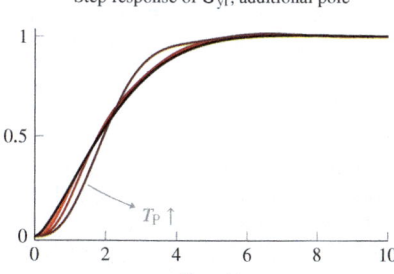 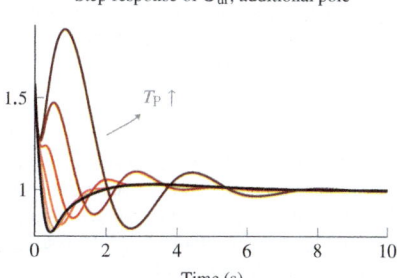

Fig. 5.16 Effect of adding an additional pole with a time constant T_P on the closed-loop step responses of controller output u and output y of an otherwise normalized plant $P(s) = \frac{1}{s^2+2s+1}$ (nominal case without additional pole: thick black line). Controller parameters remain fixed. Values of T_P are logarithmically spaced, ranging from $T_P = 0.01$ s to $T_P = 1$ s (red colors, increasingly dark)

- *Additional zero:* Secondly, the presence of an additional zero in the plant dynamics shall be examined. $P(s)$ now reads

$$P(s) = \frac{1}{s^2 + 2s + 1} \cdot \frac{T_Z s + 1}{1} = \frac{T_Z s + 1}{s^2 + 2s + 1}.$$

Again, the zero assumed to be unknown, the plant continues to be modeled with $N = 2$ and $b_0 = 1$. In Fig. 5.17, the value of T_Z is being varied between -0.2 s and 0.2 s, i.e., in a range of $\pm 20\%$ of the plant time constant T. This covers zeros in both the left and the right half of the s-plane (LHP and RHP). The latter case, found in practice in non-minimum phase systems such as certain DC-DC converter topologies, can be recognized from the characteristic undershoot in the step response of the controlled variable. RHP zeros severely restrict the achievable bandwidth of a control system, and the effects seen in Fig. 5.17 are no exception to that. The control loop with a fixed controller design is much more tolerant to LHP than RHP zeros.

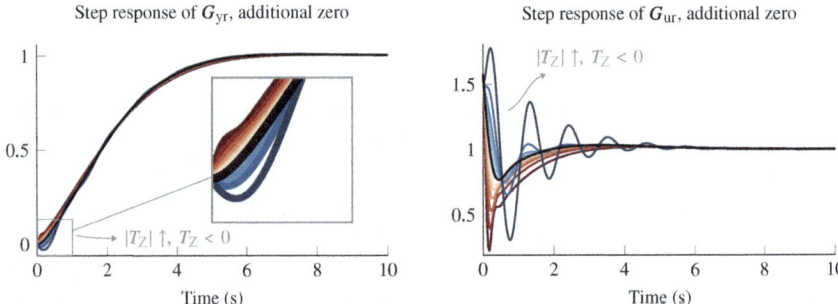

Fig. 5.17 Effect of adding an additional zero with a time constant T_Z on the closed-loop step responses of controller output u and the output y of an otherwise normalized plant $P(s) = \frac{1}{s^2+2s+1}$ (nominal case without additional zero: thick black line). Controller parameters remain fixed. Values of T_Z are logarithmically spaced, ranging from $|T_Z| = 0.02$ s (brighter colors) to $|T_Z| = 0.2$ s (darker colors) for both right-half plane (RHP, $T_Z < 0$, blue colors) and left-half plane zeros (LHP, $T_Z > 0$, red colors)

Open Access This chapter is licensed under the terms of the Creative Commons Attribution 4.0 International License (http://creativecommons.org/licenses/by/4.0/), which permits use, sharing, adaptation, distribution and reproduction in any medium or format, as long as you give appropriate credit to the original author(s) and the source, provide a link to the Creative Commons license and indicate if changes were made.

The images or other third party material in this chapter are included in the chapter's Creative Commons license, unless indicated otherwise in a credit line to the material. If material is not included in the chapter's Creative Commons license and your intended use is not permitted by statutory regulation or exceeds the permitted use, you will need to obtain permission directly from the copyright holder.

Chapter 6
Extensions and Modifications

Abstract ADRC is not a fixed set of equations and hence should not be viewed as a magical set of formulae that, once implemented on the real system, will do wonders. It should instead be understood as a modular, flexible framework that could be tailored to specific applications. This, however, requires a bit of a different mindset—one that is open to change. In this chapter, we show potential extensions and modifications to the ADRC scheme which take into account practical scenarios where certain actionable information is available about the governed system and/or acting disturbance. We also show how the performance of ADRC can be increased, for the price of additional tuning, by incorporating some nonlinear components. We will also introduce a so-called error-based variant of the ADRC scheme which is considered a bare-bones version and has some interesting practical characteristics.

6.1 Availability of Additional Model Information

The simplicity of the integrator chain plant model (3.2) assumed so far is a key strength of ADRC, taking away most of the modeling burden from the user. However, additional model information might already be available from domain knowledge, previous identification runs, data sheets, etc. Incorporating this knowledge into the observer may deliver better results, as there are fewer uncertainties the observer must deal with—with a model, one can more precisely estimate the system states, instead of moving modeling uncertainties to the lumped *total disturbance*. Similarly one can benefit from a customized disturbance generator model if the assumption of a constant disturbance does not hold for an application.

Throughout the remainder of this book, we refrain from using a more detailed model of either plant or disturbance, to keep the modeling requirements at a minimum and position ADRC as a general-purpose controller. Yet in this section, we will demonstrate how the incorporation of additional model information can be achieved by following and adapting the generic state-space derivation of ADRC presented in

Sect. 4.1, should the need ever arise. The resulting controller will have the structure shown in Fig. 4.1, and the following steps have to be carried out:

1. Provide a model of the plant and the disturbance generator in the following form:

$$\dot{x}(t) = A \cdot x(t) + b \cdot u(t) + e \cdot d(t), \quad y(t) = c^T \cdot x(t),$$
$$\dot{x}_d(t) = A_d \cdot x_d(t), \quad d(t) = c_d^T \cdot x_d(t).$$

2. Insert these matrices and vectors in the observer (4.4), whose dynamics are repeated here for convenience:

$$\begin{pmatrix} \dot{\hat{x}}(t) \\ \dot{\hat{x}}_d(t) \end{pmatrix} = \underbrace{\begin{pmatrix} A & e \cdot c_d^T \\ 0 & A_d \end{pmatrix}}_{A_o} \cdot \begin{pmatrix} \hat{x}(t) \\ \hat{x}_d(t) \end{pmatrix} + \underbrace{\begin{pmatrix} b \\ 0 \end{pmatrix}}_{b_o} \cdot u(t) + k_o \cdot \left(y(t) - \underbrace{(c^T \ 0)}_{c_o^T} \cdot \begin{pmatrix} \hat{x}(t) \\ \hat{x}_d(t) \end{pmatrix} \right).$$

The control law can now be built upon that, combining feedback of the estimated plant states and the states of the disturbance model:

$$u(t) = k_s \cdot r(t) - k_c^T \cdot \hat{x}(t) - k_d^T \cdot \hat{x}_d(t).$$

3. Design the feedback gain k_d^T of the disturbance compensation to minimize the impact of the disturbance on the plant states. Full disturbance rejection is possible if $b \cdot k_d^T = e \cdot c_d^T$ can be achieved and the observer dynamics are fast enough.
4. Compute the gains of the outer state-feedback controller k_c^T to move the poles of the plant as desired for the closed-loop dynamics, which are determined by the eigenvalues of $(A - b k_c^T)$. The *bandwidth parameterization* approach, described in Sect. 3.2.1, can be used here, of course.
5. Compute the observer gains k_o to move the observer poles, which are determined by the eigenvalues of $(A_o - k_o c_o^T)$, to the left of the closed-loop poles as desired and admissible by the application. *Bandwidth parameterization* may again be employed here to simplify pole placement.
6. Compute the static gain k_s to compensate the closed-loop gain using (4.6), which is once more repeated here for convenience:

$$k_s = - \left[c^T \cdot \left(A - b k_c^T \right)^{-1} \cdot b \right]^{-1}.$$

It should be noted that incorporating additional model information will always lead to custom controller and/or observer gain values. This means that the results and coefficients provided in the reference Chap. A of the Appendix will, in general, not be valid anymore, and the specific gains would have to be derived again, which increases the effort required to implement ADRC. If the additional performance obtained by this approach is worth, the effort should be decided in a case-by-case examination. To give some first hints, one example incorporating a more detailed plant model will be provided in the following, as well as another example with a customized disturbance model.

Example: Using Additional Plant Model Information

Let us examine a simple yet generic example of a plant with Nth-order low-pass behavior, whose transfer function $P(s)$ is

$$P(s) = \frac{y(s)}{u(s)} = \frac{b_0}{s^N + a_{N-1}s^{N-1} + \ldots + a_1 s + a_0}.$$

A state-space description of this plant can be given using the following matrix and vectors:

$$A = \begin{pmatrix} \mathbf{0}^{(N-1)\times 1} & \mathbf{I}^{(N-1)\times(N-1)} \\ -a_0 & -a_1 \cdots -a_{N-1} \end{pmatrix}, \quad b = \begin{pmatrix} \mathbf{0}^{(N-1)\times 1} \\ b_0 \end{pmatrix}, \quad c^{\mathrm{T}} = \begin{pmatrix} 1 & \mathbf{0}^{1\times(N-1)} \end{pmatrix}.$$

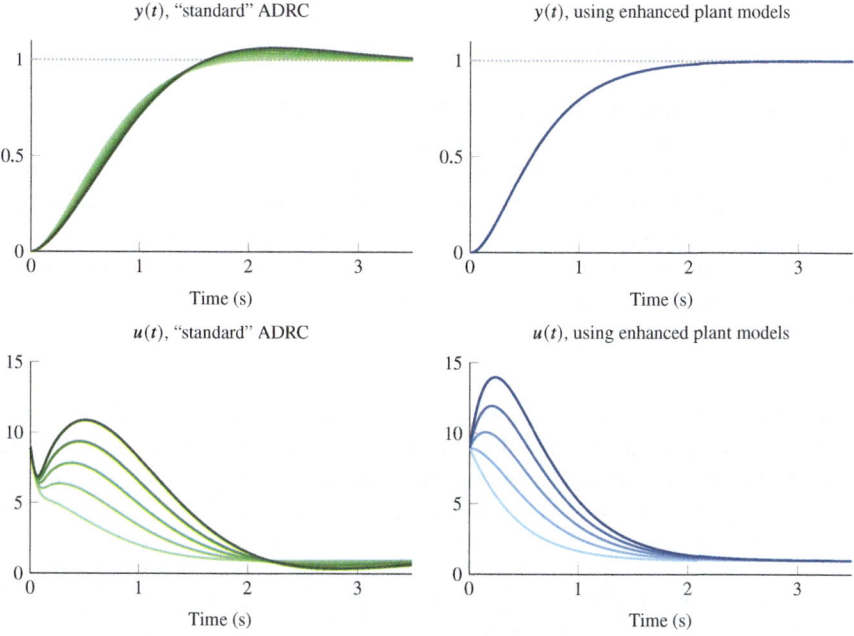

Fig. 6.1 Comparison of closed-loop step responses of ADRC without and with additional plant information in the observer. The plant has a normalized second-order low-pass behavior $P(s) = \frac{1}{s^2+2\zeta s+1}$ with variable damping ζ, which means that $b_0 = 1$ and the additional parameters of the enhanced model are $a_0 = 1$ and $a_1 = 2\zeta$. Other ADRC parameters are $N = 2$, $\omega_{\mathrm{CL}} = 3\,\mathrm{rad/s}$, and $k_{\mathrm{ESO}} = 10$. Variations of the damping range from $\zeta = 2$ (brighter) to $\zeta = 6$ (darker colors). The right-hand side results demonstrate that perfect knowledge of the plant dynamics of course helps to maintain the designed step response, whereas using "standard" ADRC will lead to certain deviations from the nominal response, with increasing lag and overshoot—albeit at a very limited scale in this example

Comparing these values with (4.8) for "standard" ADRC (as introduced in Chap. 3) reveals that additional model information for this particular plant is in the a coefficients in the Nth row of the A matrix. We leave the assumption of a constant disturbance at the plant input unchanged, which means we maintain the disturbance generator model (4.7) and gain vector (4.9):

$$\dot{x}_d(t) = \underbrace{(0)}_{A_d} \cdot x_d(t), \quad d(t) = \underbrace{(1)}_{c_d^T} \cdot x_d(t), \quad \text{and} \quad e = \begin{pmatrix} 0^{(N-1)\times 1} \\ 1 \end{pmatrix}.$$

Following the steps outlined before, we arrive at controller and observer gains k_c^T, k_o, a compensating gain k_s and a disturbance compensation gain k_d (which remains unchanged at $k_d = \frac{1}{b_0}$ compared to (4.13)). A numerical example for a second-order plant is presented in Fig. 6.1, with a comparison to "standard" ADRC.

Example: Using Additional Disturbance Model Information

In a second example, we will keep ADRC's default plant model but enhance the observer with a disturbance generator model that enables rejection of both constant and sinusoidal disturbances (with one known, fixed frequency) at the plant input. The goal here is to aid the observer with knowledge about the acting disturbance to increase its estimation performance. The disturbance $d(t)$ accordingly consists of a constant component d_0 and a sinusoidal component with unknown amplitude d_1, unknown phase shift φ_d, but known frequency ω_d:

$$d(t) = d_0 + d_1 \cdot \sin(\omega_d \cdot t + \varphi_d).$$

Generating this disturbance $d(t)$ requires superposition of an unknown constant value (as before in ADRC) and the output of a harmonic oscillator. As probably known from linear systems theory, oscillation requires a second-order system. One possible state-space model can be given as

$$\dot{x}_d(t) = \underbrace{\begin{pmatrix} 0 & 0 & 0 \\ 0 & 0 & -\omega_d^2 \\ 0 & 1 & 0 \end{pmatrix}}_{A_d} \cdot x_d(t), \quad d(t) = \underbrace{(1 \; 1 \; 0)}_{c_d^T} \cdot x_d(t).$$

The extended state observer will therefore estimate two additional state variables compared to the "standard" case, and two more observer gains have to be computed—which, of course, is not harder than before when sticking to *bandwidth parameterization* for simplicity.

Since $e = \begin{pmatrix} 0^{1\times(N-1)} & 1 \end{pmatrix}^T$ and $b = \begin{pmatrix} 0^{1\times(N-1)} & b_0 \end{pmatrix}^T$ remain unchanged assuming an Nth-order system, the condition for disturbance rejection $b \cdot k_d^T = e \cdot c_d^T$ can easily be solved for the feedback gains k_d^T, yielding

6.1 Availability of Additional Model Information

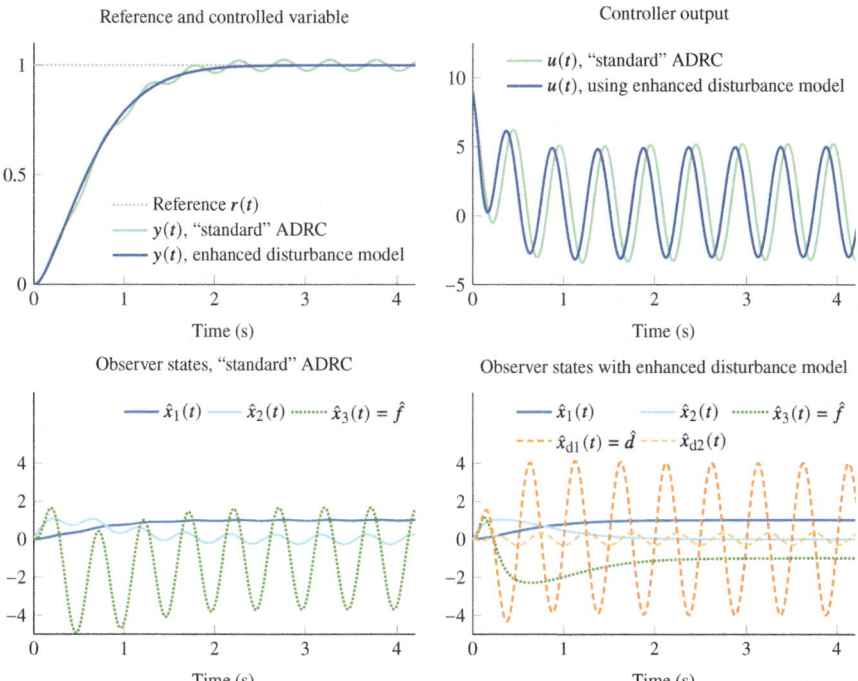

Fig. 6.2 Closed-loop step responses of "standard" ADRC and a variant with an enhanced disturbance generator model, designed to compensate a sinusoidal disturbance component with a known, fixed frequency at the plant input. The plant has a normalized second-order low-pass behavior $P(s) = \frac{1}{s^2+2s+1}$. The sinusoidal disturbance to be compensated is $d(t) = 4 \cdot \sin(\frac{2\pi}{0.5} \text{ rad/s} \cdot t)$. ADRC and tuning parameters are $N = 2$, $\omega_{\text{CL}} = 3$ rad/s, and $k_{\text{ESO}} = 10$. As visible from $y(t)$, but especially from the control action $u(t)$, even "standard" ADRC pretty well counteracts the disturbance. Only with a customized disturbance generator model, however, full rejection is possible

$$\boldsymbol{k}_{\text{d}}^{\text{T}} = \frac{1}{b_0} \cdot \boldsymbol{c}_{\text{d}}^{\text{T}} = \frac{1}{b_0} \cdot \begin{pmatrix} 1 & 1 & 0 \end{pmatrix}.$$

The "virtual plant" to be controlled by the outer state-feedback controller (introduced in Sect. 3.2.1) does not change owing to successful disturbance rejection. Computing the feedback gains $\boldsymbol{k}_{\text{c}}^{\text{T}} = \frac{1}{b_0} \cdot (k_1 \ldots k_n)$ therefore does not require changes and can, for example, directly employ the results of *bandwidth parameterization* given with (3.19). For the same reason, k_{S} does not need a customized design and remains unchanged at $k_{\text{S}} = \frac{k_1}{b_0}$.

A numerical example for a second-order plant is presented in Fig. 6.2 and compares the disturbance rejection capabilities of "standard" ADRC to the enhanced variant discussed here. As visible, a sinusoidal disturbance at the plant input can completely be removed in the latter case, if its frequency is fixed and known. It is interesting to compare the observer states: In the "standard" ADRC, all the burden is on the *total disturbance* (state $\hat{x}_3$), whereas using the enhanced disturbance model, $\hat{x}_3$ only has to account for the steady-state controller output needed to follow the reference signal. The sinusoidal disturbance is completely accounted for in the additional states $\hat{x}_{d1/2}$.

6.2 Availability of Reference Signal Derivatives

Especially in motion control systems, generating a smooth trajectory for the reference signal so as not to violate constraints regarding its velocity, acceleration or even jerk is a common approach. In such scenarios, the tracking performance of the control loop is important. So far, we have tuned ADRC for a non-overshooting response with a certain bandwidth (or settling time, e.g., using the approximation (3.21)). This is a desirable behavior for fixed set points or reference values that are (infrequently) changing in a stepwise manner—but too slow for a time-varying reference signal.

ADRC with Modified Control Law

The tracking performance of Nth-order ADRC can be considerably improved if the first N derivatives of the reference signal $r(t)$ are available. These could either be computed by a trajectory generator or estimated from a scalar reference signal (more on that in Sect. 6.3). If we put these along with the reference $r(t)$ in a reference signal vector $\boldsymbol{r}(t)$,

$$\boldsymbol{r}(t) = \begin{pmatrix} r(t) & \dot{r}(t) & \cdots & r^{(N)}(t) \end{pmatrix}^{\mathrm{T}}, \tag{6.1}$$

we can extend the control law (3.15) of ADRC as follows:

$$\begin{aligned} u(t) &= \frac{1}{b_0} \cdot \left[(\boldsymbol{k}^{\mathrm{T}} \ 1) \cdot (\boldsymbol{r}(t) - \hat{\boldsymbol{x}}(t)) \right] \\ &= \frac{1}{b_0} \cdot \left[k_1 \cdot (r(t) - \hat{x}_1(t)) + \cdots + \left(r^{(N)}(t) - \hat{x}_{N+1}(t) \right) \right]. \end{aligned} \tag{6.2}$$

Figure 6.3 shows an accordingly modified state-space variant of ADRC, with a trajectory generator providing the reference signal vector $\boldsymbol{r}(t)$ in this example. For ADRC of order N, it might not always be the case that all N derivatives of $r(t)$ are provided by the trajectory generator. In this case, using "whatever is available" may still lead to a somewhat improved tracking behavior and settling time, as demonstrated with a simulation example in Fig. 6.4.

Modular Solution

What if an implementation of ADRC is being chosen that does not allow for a modification of the state-feedback control law? As we saw in Sect. 4.2, the controller gains $\boldsymbol{k}^{\mathrm{T}}$ can become somewhat inaccessible, buried in the transfer function coefficients.

6.2 Availability of Reference Signal Derivatives

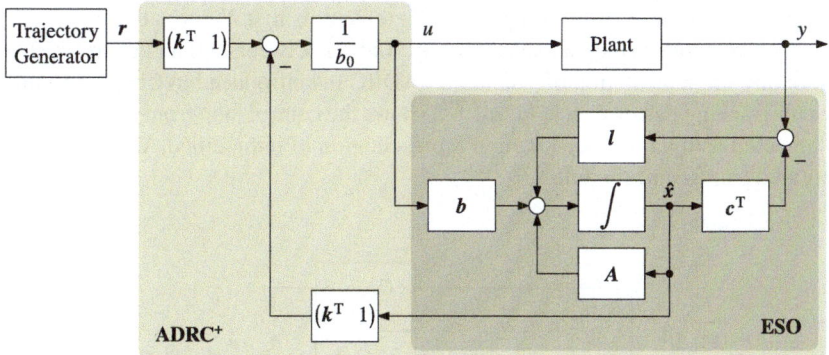

Fig. 6.3 An extended state-space variant of ADRC (accordingly labeled ADRC$^+$) making use of modified control law (6.2) to improve the tracking performance based on the availability of reference signal derivatives as part of the vector $r(t)$. The latter is in this example assumed to be provided by a dedicated trajectory generator—a common practice in motion control

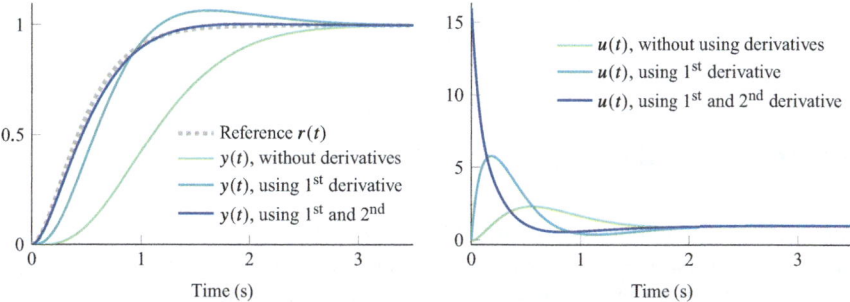

Fig. 6.4 Simulation example comparing unmodified ("standard") second-order ADRC from Chap. 3 to variants using the modified control law (6.2) with one or two derivatives of $r(t)$, which are generated by a second-order low-pass filter acting as trajectory generator. The plant is a normalized second-order system $P(s) = \frac{1}{s^2+2s+1}$. The ADRC parameters are $N = 2$, $b_0 = 1$, $\omega_{\text{CL}} = 3$ rad/s, $k_{\text{ESO}} = 10$. Clearly using the reference signal derivatives significantly improves the tracking performance, with almost ideal tracking when using all ($N = 2$) derivatives—if the enlarged controller output values $u(t)$ are feasible in the actual application

Fortunately, by comparing Figs. 6.3 and 3.5, we can derive a modular solution. If we factor out the gain vector $\frac{1}{k_1} \cdot \left(k^\text{T} \ 1 \right)$ from the reference signal input of the controller in Fig. 6.3, we obtain a series connection of said gain vector and an unmodified ADRC structure of Fig. 3.5—which can, of course, be implemented using transfer functions, as well. In terms of the control law (6.2), this means

$$u(t) = \frac{1}{b_0} \cdot \left[\left(k^\text{T} \ 1 \right) \cdot (r(t) - \hat{x}(t)) \right] = \frac{1}{b_0} \cdot \left[\left(k^\text{T} \ 1 \right) \cdot r(t) - \left(k^\text{T} \ 1 \right) \cdot \hat{x}(t) \right]$$

$$= \frac{1}{b_0} \cdot \left[k_1 \cdot \underbrace{\left(\frac{1}{k_1} \cdot \left(k^\text{T} \ 1 \right) \cdot r(t) \right)}_{\text{input to "standard" ADRC}} - \left(k^\text{T} \ 1 \right) \cdot \hat{x}(t) \right].$$

If a reference vector (6.1), comprising of $r(t)$ and its first N derivatives, is available, one can therefore plug a gain vector $\frac{1}{k_1} \cdot \begin{pmatrix} k^T & 1 \end{pmatrix}$ between $r(t)$ and the reference input of any implementation of "standard" ADRC (as introduced in Chap. 3) to improve the tracking performance. Figure 6.5 shows the control loop from Fig. 6.3, now implemented using a transfer function representation of unmodified ADRC and the mentioned gain vector module.

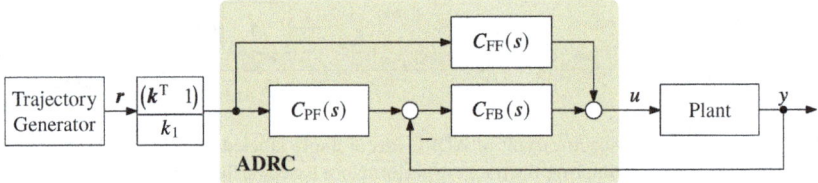

Fig. 6.5 Control loop with "standard" ADRC (as understood in this book), here in a transfer function implementation from Sect. 4.2.2, enhanced by a gain vector module $\frac{1}{k_1} \cdot \begin{pmatrix} k^T & 1 \end{pmatrix}$ at the reference signal input to improve tracking performance if reference derivatives are available as part of the reference vector $r(t)$

6.3 Nonlinear ADRC

Aiming at design simplicity and clarity of presentation, we made a conscious decision early on in the book to introduce and explain ADRC in its linear form. However one should be aware that incorporating some nonlinear components in the ADRC structure can potentially improve its performance, even for linear plants, although nonlinear ADRC (often denoted as NADRC) is not common in engineering practice. Interestingly, it is a foundational part of ADRC's history: The original ADRC version was introduced as a nonlinear scheme. We will provide more context in Chap. 7. Here we show how the inclusion of nonlinear terms can be done using a specific form of NADRC that has two distinct elements that make it nonlinear, namely:

- A nonlinear *tracking differentiator*, responsible for real-time estimation of signal's higher-order derivatives
- A nonlinear weighting function, which "scales" selected signals in observer and/or controller within the ADRC structure to improve their convergence rate

From a practical standpoint, if there exists a justifiable need to increase the performance offered by the linear ADRC, these nonlinear components can be added to the core design (see Fig. 6.6) and, for the price of more complicated structure, additional tuning, and increased computational complexity, can improve the desired aspect of ADRC, like tracking and/or observation quality. Next, we will show how the above nonlinear elements can be incorporated into the base linear ADRC.

6.3 Nonlinear ADRC

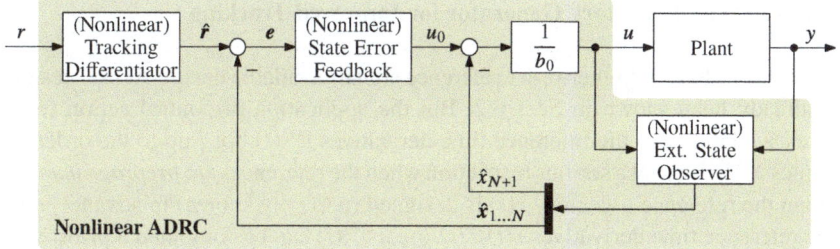

Fig. 6.6 Closed control loop with nonlinear ADRC, consisting of nonlinear tracking differentiator (TD), nonlinear extended state observer (NESO), and the outer controller in the form of a nonlinear feedback of the state error (NLSEF)

6.3.1 Tracking Differentiator

The tracking differentiator, or TD, can generate estimates of a signal and its higher-order derivatives. Using reference signal $r(t)$ as an exemplary input, the TD can be expressed as the following second-order integral chain:

$$\ddot{\hat{r}}(t) = -\rho \cdot \text{sign}\left(\hat{r}(t) - r(t) + \frac{\dot{\hat{r}}(t) \cdot |\dot{\hat{r}}(t)|}{2 \cdot \rho}\right), \quad (6.3)$$

$$\dot{\hat{r}}(t) = \int \ddot{\hat{r}}(t)\, dt,$$

$$\hat{r}(t) = \int \dot{\hat{r}}(t)\, dt,$$

where $\hat{r}(t)$, $\dot{\hat{r}}(t)$, $\ddot{\hat{r}}(t)$ are to be used as estimates of $r(t)$, $\dot{r}(t)$, $\ddot{r}(t)$, respectively, and $\rho > 0$ is the design parameter that balances speed of estimation and stability of the estimator. Figure 6.7 shows the TD in block diagram form.

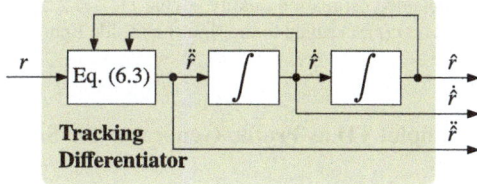

Fig. 6.7 Block diagram of the TD implementation based on (6.3), providing an estimation of the input signal and its first two derivatives

The information about the estimated higher-order terms of the reference signal can be utilized in various ways. The following two examples show its usage as a trajectory generator and a transient profile generator. Both examples use the same TD structure (6.3) but differ in tuning.

Example: TD as Trajectory Generator for Improved Tracking

The benefits of having higher-order reference signals available during ADRC design have already been shown in Sect. 6.2. But the application of control action (6.2) requires availability of the reference time derivatives $r^{(i)}(t)$, for i up to the order of dynamics N. This is not a serious restriction when the references are *preprogrammed*, meaning the reference trajectory $r(t)$ is designed (perfectly known) in advance, and all the reference time derivatives $\dot{r}(t), \ddot{r}(t), \ldots, r^{(N)}(t)$ can be computed a priori and fed to the controller.

In some control applications, however, the value of the reference trajectory is only known at a current time instant and the reference time derivatives are unavailable. In those instances, TD can be possibly used to reconstruct the otherwise unavailable signals in real time to provide feedforward information improving command response, similar to the trajectory generator depicted in Fig. 6.5. Using a generic sine signal as an example, Fig. 6.8 shows a potential use of a TD as the trajectory generator.[1] In this case, the signals estimated by TD should be as close as possible to the real ones; hence tuning coefficient ρ needs to be set to a relatively large value.

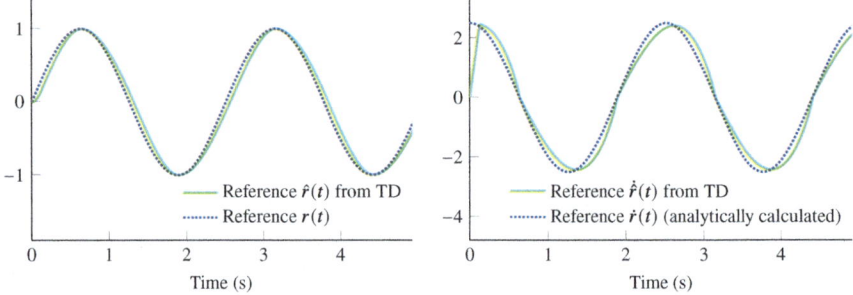

Fig. 6.8 The use of TD (6.3) as a source of higher-order reference signal in the trajectory generator. The TD is responsible here for reconstructing in real time the estimate $\dot{\hat{r}}(t)$ based on signal $r(t) = \sin(2.5 \cdot t)$. Signal $\hat{r}(t)$ is also estimated as a by-product of implementing (6.3). To compare the obtained estimation quality, signal $\dot{r}(t) = 2.5\cos(2.5 \cdot t)$, which was calculated analytically based on $r(t)$, is added to the right-hand side figure. The TD was tuned with $\rho = 20$

Example: TD as Profile Generator for Smooth Transients

Another possible use of TD is that if the given set point is abrupt, then the TD estimates could provide a smooth transient profile that the output of the plant can then reasonably follow, which should be assessed on a case-by-case basis. Such avoidance of set point jumps is oftentimes a standard operating procedure when dealing, for

---

[1] Another interesting approach addressing the issue of unavailability of higher-order terms of the reference signal is the use of so-called *error-based* ADRC, which we will cover later in Sect. 6.4.

6.3 Nonlinear ADRC

example, with servo systems, for which motion profiles should be devised to ensure safe operation and not to overstrain the hardware components. In that case, if $r(t)$ is the desired servo position, then TD in form of (6.3) would provide motion profiles for its reference position $\hat{r}(t)$, velocity $\dot{\hat{r}}(t)$, and acceleration $\ddot{\hat{r}}(t)$, to be then utilized during controller synthesis.

Using a generic step signal as an example, Fig. 6.9 shows a potential use of a TD as the transient profile generator. In this case, the signals estimated by TD do not need to be as close as possible to the real signals as the former are meant to include the feasibility of their practical realization; hence tuning coefficient ρ needs to be set to a relatively small value.

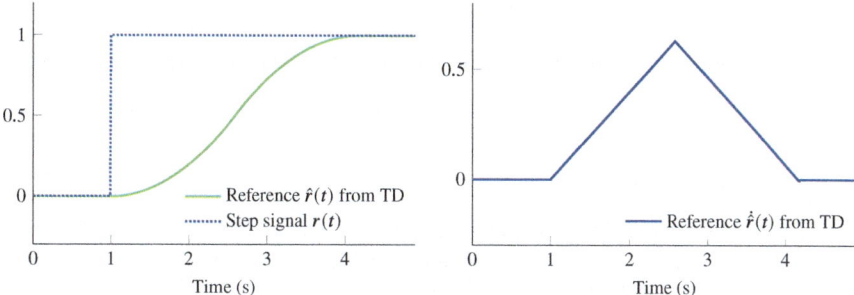

Fig. 6.9 Using TD (6.3) as a higher-order for transient profile generator for reference signals. The TD is responsible here for reconstructing in real-time transient, practically realizable shapes of $\hat{r}(t)$ and $\dot{\hat{r}}(t)$, based solely on an input signal $r(t)$. The TD was tuned with $\rho = 0.4$

6.3.2 Nonlinear Controller and Observer

Until now, when constructing ADRC, we have utilized a simple, linear combination of gains and signals in both the observer part and the controller part. However, a nonlinear combination may potentially be more effective, if one is willing to put in the extra work related to the implementation and tuning of additional components. Here we will show how to take advantage of nonlinear weighting in ADRC, starting by adding it to the baseline linear controller, discussed so far in previous chapters.

From Linear to Nonlinear Controller

To convey the idea behind a nonlinear controller, let us use as an example the specific form of ADRC discussed in Sect. 6.2. This time, however, we assume that the reference signal derivatives are not available beforehand but can be estimated in real time based solely on the known reference signal $r(t)$, for example, using a TD. This assumption means that in the extended control law (6.2), we can now replace $r(t)$ with its estimate:

$$\hat{\boldsymbol{r}}(t) = \left(\hat{r}(t) \ \dot{\hat{r}}(t) \ \cdots \ \hat{r}^{(N)}(t)\right)^{\mathrm{T}} = \left(\hat{r}_1(t) \ \hat{r}_2(t) \ \cdots \ \hat{r}_{N+1}(t)\right)^{\mathrm{T}}, \qquad (6.4)$$

which yields the following modified version of (6.2):

$$u(t) = \frac{1}{b_0} \cdot \left[\left(\boldsymbol{k}^{\mathrm{T}} \ 1\right) \cdot \left(\hat{\boldsymbol{r}}(t) - \hat{\boldsymbol{x}}(t)\right) \right]. \qquad (6.5)$$

Furthermore, if we introduce the following error vector:

$$\boldsymbol{e}(t) = \begin{pmatrix} e_1 \\ \vdots \\ e_N \end{pmatrix} = \begin{pmatrix} \hat{r}_1(t) - \hat{x}_1(t) \\ \vdots \\ \hat{r}_N(t) - \hat{x}_N(t) \end{pmatrix}, \qquad (6.6)$$

the state-feedback controller $u_0(t)$, inside (6.5), can be written in a compact form:

$$u_0(t) = \overbrace{\boldsymbol{k}^{\mathrm{T}} \cdot \boldsymbol{e}(t)}^{\text{"standard" linear weighting}}, \qquad (6.7)$$

with distinct linear weighting, which is emblematic of the "standard" LADRC form discussed so far in the book. Now, the above simple linear combination could also be replaced with a nonlinear one, thus changing (6.7) to

$$u_0(t) = \overbrace{\boldsymbol{k}^{\mathrm{T}} \cdot \boldsymbol{f}\left(\boldsymbol{e}(t)\right)}^{\text{nonlinear weighting}}, \qquad (6.8)$$

where

$$\boldsymbol{f}\left(\boldsymbol{e}(t)\right) = \begin{pmatrix} f(e_1(t), \boldsymbol{p}_1) \\ \vdots \\ f(e_N(t), \boldsymbol{p}_N) \end{pmatrix}^{\mathrm{T}} \qquad (6.9)$$

is some function that scales the entering signal in some nonlinear fashion with $\boldsymbol{p}_1, \ldots, \boldsymbol{p}_N$ being user-defined design parameters that shape this nonlinear behavior. The control law (6.8) with such nonlinear terms is sometimes referred to in the ADRC literature as *nonlinear state error feedback* or NLSEF. A specific example of such a nonlinear function will be shown later on.

From Linear to Nonlinear Observer

A similar transition from linear to nonlinear, as seen above for the controller part, can be done for the observer part. With the following auxiliary definition of estimation error, here being the difference between measured output and its estimate:

$$e_y(t) = y(t) - \boldsymbol{c}^{\mathrm{T}} \cdot \hat{\boldsymbol{x}}(t), \qquad (6.10)$$

6.3 Nonlinear ADRC

let us recall the linear ESO from (3.16):

$$\dot{\hat{x}}(t) = A \cdot \hat{x}(t) + b \cdot u(t) + \underbrace{l \cdot e_y(t)}_{\text{"standard" linear weighting}}, \tag{6.11}$$

with the marked characteristic linear weighting between signal and gains. Here again, the linear weighting could be replaced with a nonlinear one, thus changing (6.11) to

$$\dot{\hat{x}}(t) = A \cdot \hat{x}(t) + b \cdot u(t) + \underbrace{l \cdot f\left(e_y(t)\right)}_{\text{nonlinear weighting}}, \tag{6.12}$$

where

$$f\left(e_y(t)\right) = \begin{pmatrix} f(e_y(t), p_1) \\ \vdots \\ f(e_y(t), p_{N+1}) \end{pmatrix} \tag{6.13}$$

is a nonlinear function that scales the entering signal, with $p_1, \ldots, p_{N+1}$ being the design parameters. The observer (6.12) with such nonlinear terms is often referred to in the literature on ADRC simply as *nonlinear ESO* or NESO. Let us now take a look at a specific example of the abovementioned nonlinear weighting function.

Specific Case of Nonlinear Weighting Function

In general, several different weighting functions can be used in nonlinear ADRC to replace the generic functions (6.9) and (6.13) in the controller and the observer, respectively. Here, however, we are focusing on a specific one, commonly referred to as $f_{\text{al}}(\cdot)$ function:[2]

$$x_{\text{out}}(t) = f_{\text{al}}(x_{\text{in}}(t), \alpha, \delta) = \begin{cases} \dfrac{x_{\text{in}}(t)}{\delta^{(1-\alpha)}}, & |x_{\text{in}}(t)| \leq \delta, \\ |x_{\text{in}}(t)|^\alpha \cdot \text{sign}(x_{\text{in}}(t)), & |x_{\text{in}}(t)| > \delta, \end{cases} \tag{6.14}$$

where $x_{\text{in}}(t)$ is some signal entering the function and $x_{\text{out}}(t)$ is its weighted output that this nonlinear function produces. A graphical example of how the $f_{\text{al}}(\cdot)$ function does nonlinear weighting is depicted in Fig. 6.10.

Both quantity and placement of the nonlinear functions within the constructed ADRC scheme are fully customizable by the control designer, as will be shown in the upcoming example. This decision, which could increase the convergence rate of selected signals, has to be made taking into account its practical feasibility, as it will have consequences on the overall implementation complexity and require additional tuning effort.

---

[2] Here again, there is a special historical meaning behind discussing this particular function, which we will explain in Chap. 7.

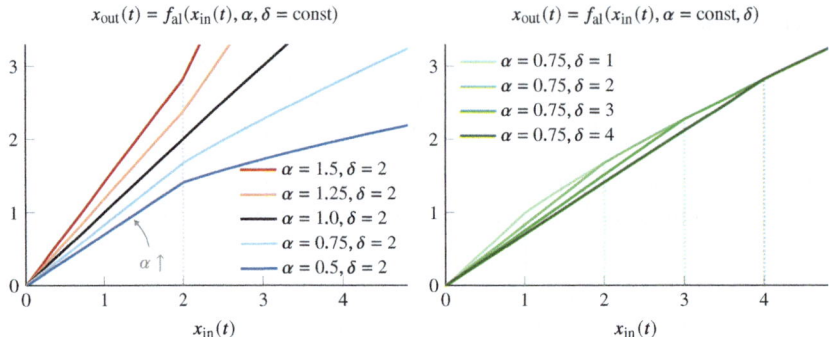

Fig. 6.10 Influence of the nonlinear $f_{\text{al}}(\cdot)$ function design parameters α (left; depicted for common $\delta = 2$) and δ (right; depicted for common $\alpha = 0.75$). Parameter δ represents the value of the input signal at which the instantaneous scaling occurs and α represents the *intensity* of said scaling. In a special case when $\alpha = 1$, nonlinear scaling does not occur at all (regardless of the value of δ), and $f_{\text{al}}(\cdot)$ function has no effect on the input signal, namely $x_{\text{out}}(t) = x_{\text{in}}(t)$ (black graph on the left)

Example: Influence of Nonlinear Weighting

Through the following simulation example, we want to show what is the influence of the nonlinear components on the ADRC performance. Let us examine once again the nominal and generic second-order plant model from Sect. 5.1 to compare two ADRC designs, one with "standard" linear weighting and the other with nonlinear weighting using $f_{\text{al}}(\cdot)$ function.

In the case of linear controller, we have

$$u(t) = \frac{1}{b_0} \cdot \overbrace{\left(k_1 \cdot e_1(t) + k_2 \cdot e_2(t)\right.}^{\text{"standard" linear weighting}} \underbrace{-\hat{x}_3(t)}_{\text{"standard" linear weighting}}) \tag{6.15}$$

and in the case of a nonlinear controller

$$u(t) = \frac{1}{b_0} \cdot \overbrace{\left(k_1 \cdot f_{\text{al}}(e_1(t), \alpha_1, \delta) + k_2 \cdot f_{\text{al}}(e_2(t), \alpha_2, \delta)\right.}^{\text{nonlinear weighting}} \underbrace{-\hat{x}_3(t)}_{\text{nonlinear weighting}}). \tag{6.16}$$

In both cases, we use the definitions of error signals $e_1(t) = \hat{r}_1(t) - \hat{x}_1(t)$ and $e_2(t) = \hat{r}_2(t) - \hat{x}_2(t)$. For the NLSEF (6.16), we make an arbitrary decision to place the nonlinear weighting on both feedback error-driven terms. Estimates $\hat{r}_1(t)$ and $\hat{r}_2(t)$ are taken from the same TD (similar to (6.3)), whereas estimates $\hat{x}_1(t)$ and $\hat{x}_2(t)$ are taken from respective observers.

Now, for the observer part in the tested ADRCs, the purely linear observer takes the "standard" form:

$$\underbrace{\begin{pmatrix}\dot{\hat{x}}_1(t)\\ \dot{\hat{x}}_2(t)\\ \dot{\hat{x}}_3(t)\end{pmatrix}}_{\dot{\hat{x}}(t)} = \underbrace{\begin{pmatrix}0 & 1 & 0\\ 0 & 0 & 1\\ 0 & 0 & 0\end{pmatrix}}_{A} \cdot \underbrace{\begin{pmatrix}\hat{x}_1(t)\\ \hat{x}_2(t)\\ \hat{x}_3(t)\end{pmatrix}}_{\hat{x}(t)} + \underbrace{\begin{pmatrix}0\\ b_0\\ 0\end{pmatrix}}_{b} \cdot u(t) + \underbrace{\begin{pmatrix}l_1\\ l_2\\ l_3\end{pmatrix}}_{l} \cdot \overbrace{e_y(t)}^{\text{linear weighting}}, \quad (6.17)$$

and in case of the mixed linear and nonlinear weighting, the observer writes

$$\underbrace{\begin{pmatrix}\dot{\hat{x}}_1(t)\\ \dot{\hat{x}}_2(t)\\ \dot{\hat{x}}_3(t)\end{pmatrix}}_{\dot{\hat{x}}(t)} = \underbrace{\begin{pmatrix}0 & 1 & 0\\ 0 & 0 & 1\\ 0 & 0 & 0\end{pmatrix}}_{A} \cdot \underbrace{\begin{pmatrix}\hat{x}_1(t)\\ \hat{x}_2(t)\\ \hat{x}_3(t)\end{pmatrix}}_{\hat{x}(t)} + \underbrace{\begin{pmatrix}0\\ b_0\\ 0\end{pmatrix}}_{b} \cdot u(t) + \underbrace{\begin{pmatrix}l_1\\ l_2\\ l_3\end{pmatrix}}_{l} \cdot \overbrace{\begin{pmatrix}e_y(t)\\ e_y(t)\\ f(e_y(t), \alpha, \delta)\end{pmatrix}}^{\text{mix of linear and nonlinear weighting}}. \quad (6.18)$$

In both cases, we have $e_y(t) = x_1(t) - \hat{x}_1(t) = y(t) - \hat{y}(t)$. For the NESO (6.18), we make an arbitrary decision to place the nonlinear weighting only on the last term (estimating the *total disturbance*). This again shows how customizable the placement of nonlinear terms is. Here it results from a practical compromise between the desire to increase the performance of ADRC and its overall complexity of implementation and tuning (i.e., less nonlinear functions mean simpler tuning).

For the comparison, we start with "standard" LADRC and add nonlinear components to it, first in the observer, then in the controller, and finally in both of them. The obtained results are presented in Fig. 6.11. It can be seen that nonlinear terms can bring benefits in certain aspects but at the cost of some drawbacks in other areas. The proper selection, placement, and tuning of nonlinear extra terms should thus be a conscious decision made based on actual control needs and limitations that inevitably result in some kind of compromise. For example, in the case of structure NADRC$_3$, the nonlinear terms not only increased disturbance rejection and response speed but also resulted in a larger overshoot than in its linear counterpart. And once again, possible improvements from adding nonlinearity will come with a rather significant price tag, namely more elements need to be implemented and tuned. On a personal note, we suggest always starting with a linear ADRC and going from there. As long as the job is done, engineering practice favors simplicity. This is yet another reason this book focuses almost exclusively on linear ADRC form.

6.4 Error-Based ADRC

The previous three sections were about "adding things" to the ADRC scheme. Let us now exercise the idea of potentially simplifying it and checking what benefits does that bring. However, as the adage goes, there is no such thing as free lunch.

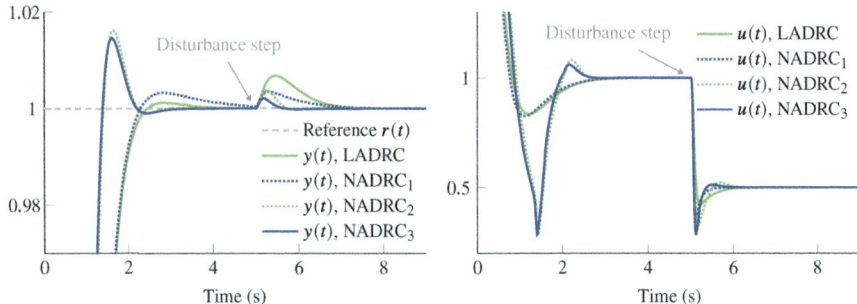

Fig. 6.11 Simulation example comparing step responses of unmodified ("standard") LADRC from Chap. 3 to its nonlinear variants, namely structures with NESO (denoted as NADRC$_1$), with NLSEF (NADRC$_2$), and with both NESO and NLSEF (NADRC$_3$). A normalized second-order system $P(s) = \frac{1}{s^2+2s+1}$ is used as the plant with ADRC parameters $N = 2$, $b_0 = 1$, $\omega_{\mathrm{CL}} = 3\,\mathrm{rad/s}$, and $k_{\mathrm{ESO}} = 10$. The $f_{\mathrm{al}}(\cdot)$ function parameters are tuned to improve disturbance rejection. In the nonlinear observer (6.18), they are chosen as $\alpha = 1.75$ and $\delta = 8 \cdot 10^{-5}$. In the nonlinear controller (6.16), they are $\alpha_1 = 0.7$, $\alpha_2 = 0.9$, and $\delta = 3 \cdot 10^{-3}$. To test disturbance rejection, an unmodeled input disturbance $d = 0.5$ is applied starting from $t = 5\,\mathrm{s}$. Figures are zoomed in for better legibility

So it will also be necessary to answer the important question about the price of said simplification.

Here we focus on a particular modification of the ADRC structure, which is referred to as *error-based* ADRC; coining of this term will become clear in its upcoming derivation. This modification is dedicated to scenarios in which trajectory tracking control tasks are considered, but, in contrast to Sect. 6.2, the consecutive reference time derivatives are not measured or generated using a trajectory generator. In other words, they are not available and hence cannot be used during the controller synthesis. We have made a conscious decision to include the error-based variant in the book as it offers some interesting, practical features that may be found useful by prospective ADRC designers, but we will get to that later.

To best explain error-based ADRC, we start by recalling the Nth-order SISO plant model (3.1), already expressed in its compact form (3.2):

$$y^{(N)}(t) = f(t) + b_0 \cdot u(t), \tag{6.19}$$

as well as the fundamental definition of feedback error:

$$e(t) = r(t) - y(t). \tag{6.20}$$

By differentiating (6.20) up to the order of governed system dynamics (6.19)

$$\dot{e}(t) = \dot{r}(t) - \dot{y}(t),$$

$$\vdots$$

$$e^{(N)}(t) = r^{(N)}(t) - y^{(N)}(t), \tag{6.21}$$

6.4 Error-Based ADRC

one reaches the point where (6.19) can be substituted in (6.21), which, after a simple rearrangement of terms, can be expressed as

$$
\begin{aligned}
e^{(N)}(t) &= r^{(N)}(t) - \overbrace{(f(t) + b_0 \cdot u(t))}^{y^{(N)}(t)} \\
&= \underbrace{r^{(N)}(t) - f(t)}_{\hat{f}^*(t)} - b_0 \cdot u(t) \\
&= f^*(t) - b_0 \cdot u(t),
\end{aligned}
\qquad (6.22)
$$

where $f^*(t)$ now is the *total disturbance*, which, compared to (6.19), now not only consists of unknown model terms (modeling errors) and actual disturbances but also $r^{(N)}(t)$ being the unknown Nth derivative of the reference signal.

One can notice that the considered plant model can be expressed in two alternative forms, (6.19) or (6.22). Two major differences can be immediately spotted between the two expressions. Firstly, (6.19) represents system output dynamics, whereas (6.22) represents its error dynamics. The second difference is the sign attached to the *critical gain parameter* b_0.

For the considered error-based system (6.22), we can follow the methodology of ADRC design established in Chap. 3 and construct a control law, similar to (3.3), that embodies the two core ideas of ADRC, namely disturbance rejection and plant gain inversion. This, of course, assumes it is possible to obtain at least an estimate $\hat{f}^*(t)$ of the *total disturbance* $f^*(t)$ in an online manner. Such control signal $u(t)$ can then be constructed as[3]

$$
u(t) = \frac{u_0(t) + \hat{f}^*(t)}{b_0}. \qquad (6.23)
$$

The application of control signal $u(t)$ to (6.22) creates the desired unity-gain integrator chain behavior $e^{(N)}(t) \approx -u_0(t)$, which is straightforward to control.

The above formulation of the control problem in the error domain is thus potentially very practically appealing as it allows the realization of the control objective without knowledge of the reference derivatives. However, as is usually the case in ADRC schemes, it all comes down to the estimation quality of the *total disturbance*. Fortunately, we can reach for known tools and methods, like ESO and *bandwidth parameterization*, to effectively reconstruct the missing information.

We see important benefits of introducing error-based ADRC using both state-space and transfer function representations; hence these two forms are derived in the remainder of this section.

---

[3] Please notice the sign change next to the *total disturbance* term, compared to (3.3).

6.4.1 State-Space Form

Similar to the output-based case in Sect. 3.1, a good starting point here is the expression of (6.22)—the disturbed integrator chain in error domain—in state-space form, which results in

$$\underbrace{\begin{pmatrix} \dot{x}_1 \\ \vdots \\ \dot{x}_{N+1} \end{pmatrix}}_{\dot{\boldsymbol{x}}(t)} = \underbrace{\begin{pmatrix} \boldsymbol{0}^{N\times 1} & \boldsymbol{I}^{N\times N} \\ 0 & \boldsymbol{0}^{1\times N} \end{pmatrix}}_{\boldsymbol{A}} \cdot \underbrace{\begin{pmatrix} x_1(t) \\ \vdots \\ x_{N+1}(t) \end{pmatrix}}_{\boldsymbol{x}(t)} - \underbrace{\begin{pmatrix} \boldsymbol{0}^{(N-1)\times 1} \\ b_0 \\ 0 \end{pmatrix}}_{\boldsymbol{b}} \cdot u(t) + \begin{pmatrix} \boldsymbol{0}^{N\times 1} \\ 1 \end{pmatrix} \cdot \dot{f}^*(t) \qquad (6.24)$$

$$e(t) = \underbrace{\begin{pmatrix} 1 & \boldsymbol{0}^{1\times N} \end{pmatrix}}_{\boldsymbol{c}^{\mathrm{T}}} \cdot \begin{pmatrix} x_1(t) \\ \vdots \\ x_{N+1}(t) \end{pmatrix}.$$

From $\boldsymbol{c}^{\mathrm{T}}$ in (6.24) it is clear that the first state variable is $x_1(t) = e(t)$. Looking at $\boldsymbol{A}$ furthermore reveals that $\dot{x}_i(t) = x_{i+1}(t)$ holds for $i = 1, \ldots, N-1$, which means that $x_{i+1}(t)$ is the ith derivative of $e(t)$. In other words, the estimated state vector $\hat{\boldsymbol{x}}(t)$ in (6.24), contrary to what has been shown earlier in (3.16), now consists of estimated feedback error $\hat{e}(t)$ and its consecutive time derivatives. The next-to-last state equation reads $\dot{x}_N(t) = x_{N+1}(t) - b_0 \cdot u(t)$, corresponding to $e^{(N)}(t) = f^*(t) - b_0 \cdot u(t)$. This means that $f^*(t)$ is the final element of the vector of state variables, $x_{N+1}(t) = f^*(t)$, and consequently, the last line correctly states $\dot{x}_{N+1}(t) = \dot{f}^*(t)$.

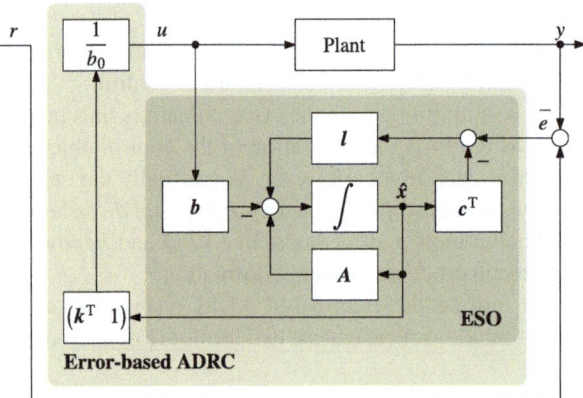

Fig. 6.12 Continuous-time error-based ADRC in state space

For the above extended state system model in state space, the following Luenberger observer can be designed to obtain an estimate of the states $\hat{\boldsymbol{x}}^{\mathrm{T}}(t) = \begin{pmatrix} \hat{x}_1(t) & \cdots & \hat{x}_{N+1}(t) \end{pmatrix}$, with $u(t)$ and $e(t)$ being the only measurements available:

6.4 Error-Based ADRC

$$\underbrace{\begin{pmatrix} \dot{\hat{x}}_1 \\ \vdots \\ \dot{\hat{x}}_{N+1} \end{pmatrix}}_{\dot{\hat{x}}(t)} = \underbrace{\begin{pmatrix} \mathbf{0}^{N\times 1} & \mathbf{I}^{N\times N} \\ 0 & \mathbf{0}^{1\times N} \end{pmatrix}}_{A} \cdot \underbrace{\begin{pmatrix} \hat{x}_1(t) \\ \vdots \\ \hat{x}_{N+1}(t) \end{pmatrix}}_{\hat{x}(t)} - \underbrace{\begin{pmatrix} \mathbf{0}^{(N-1)\times 1} \\ b_0 \\ 0 \end{pmatrix}}_{b} \cdot u(t) + \underbrace{\begin{pmatrix} l_1 \\ \vdots \\ l_{N+1} \end{pmatrix}}_{l} \cdot (e(t) - \hat{e}(t))$$

$$\hat{e}(t) = \underbrace{\begin{pmatrix} 1 & \mathbf{0}^{1\times N} \end{pmatrix}}_{c^{\mathrm{T}}} \cdot \begin{pmatrix} \hat{x}_1(t) \\ \vdots \\ \hat{x}_{N+1}(t) \end{pmatrix}.$$

(6.25)

This can be written in a more compact way as

$$\dot{\hat{x}}(t) = A \cdot \hat{x}(t) - b \cdot u(t) + l \cdot \left(e(t) - c^{\mathrm{T}} \cdot \hat{x}(t) \right), \qquad (6.26)$$

with $l = (l_1 \cdots l_{N+1})^{\mathrm{T}}$ being the observer gains, to be tuned using, for example, the already introduced *bandwidth parameterization* approach from Sect. 3.2.2.

Here, similar to (3.9), we omit the "virtual input" $\hat{f}^*(t)$, which means setting the first derivative of $\hat{f}^*(t)$ to zero, which results in $\hat{x}_{N+1}(t)$ being estimated as a constant *total disturbance* at the plant model input. That way, the extended state variable $\hat{x}_{N+1}(t)$ serves as the integrator state of ADRC necessary to achieve zero steady-state error for constant reference values and constant disturbances.

Please notice that now the control signal governing system (6.22) does not have to be designed as previously seen, for example, in (3.15), but can take advantage of the fact that the estimated state vector $\hat{x}(t)$ in observer (6.25) contains the estimated feedback error and its consecutive derivatives and thus can be constructed simply as

$$u(t) = \frac{1}{b_0} \cdot \left(k^{\mathrm{T}} \; 1 \right) \cdot \hat{x}(t) \quad \text{with} \quad k^{\mathrm{T}} = (k_1 \cdots k_N), \qquad (6.27)$$

where k is the vector of controller gains, which again can be tuned using, for example, the *bandwidth parameterization* approach as seen in Sect. 3.2.1.

Finally, having now both the observer and the controller parts, the error-based ADRC in state space can be shown in Fig. 6.12. Immediately, through a quick inspection and comparison with its output-based counterpart in Fig. 3.5, one can visually notice the two abovementioned differences between output- and error-based ADRC schemes. The first one is the fact that the input signal to the error-based ADRC scheme is not the plant output $y(t)$ but the feedback error $e(t)$, and the second one is the different sign applied to the input $b \cdot u(t)$.

So what is the justification behind expressing ADRC in such an alternative, error-based form? There are several reasons, but let us start with the obvious one. The availability of reference derivatives is no longer an issue, as they are being indirectly reconstructed on the fly by the observer, which estimates the feedback error and its consecutive derivatives. When fixed set point control tasks are considered, which means that higher-order derivatives of the reference signal are zero (except for possible stepwise changes to the set point), error-based ADRC offers a very peculiar behavior, especially when compared with the "standard" output-based variant.

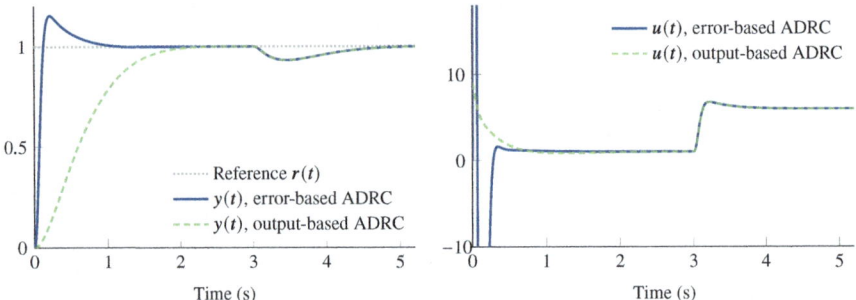

Fig. 6.13 Closed-loop step responses and disturbance rejection using error-based (solid lines) and output-based ADRC (dashed lines). The plant has a normalized second-order low-pass behavior $P(s) = \frac{1}{s^2+2s+1}$. Tuning parameters are $N = 2$, $\omega_{CL} = 3\,\text{rad/s}$, and $k_{ESO} = 10$. A disturbance $d = -5$ becomes effective at the plant input starting from $t = 3\,\text{s}$. Note that the much-reduced rise time with error-based ADRC requires giant control efforts $u(t)$, which are not fully visible here, as the $u(t)$ diagram was zoomed in to focus on the identical disturbance rejection behavior

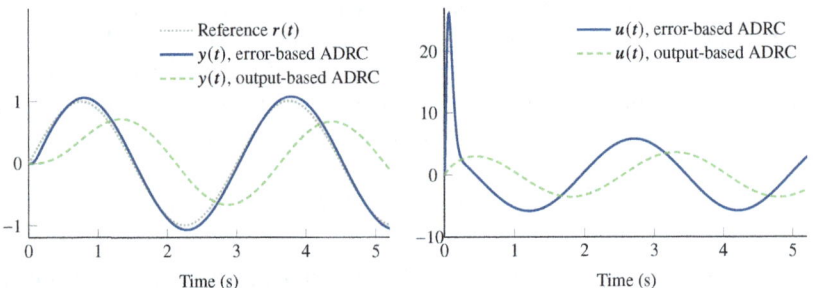

Fig. 6.14 Tracking performance of error-based (solid lines) and output-based ADRC (dashed lines) with a sinusoidal reference $r(t) = \sin(\frac{2\pi}{3\,\text{s}} \cdot t)$. The same plant and tuning parameters as in Fig. 6.13

This has been visualized in Fig. 6.13, where error-based ADRC generates a very rapid transient system response, resulting from a large initial control signal that, in return, generates a significant overshoot. This is the result of error-based ADRC estimating the unknown reference signal derivatives—which are huge for stepwise reference changes. Interestingly, both tested control variants have the same level of disturbance rejection—a characteristic we will go back to later in this section.

On the other hand, the error-based ADRC shows its potential when it is used in trajectory tracking tasks, like in the second example depicted in Fig. 6.14, in which the reference is continually changing. It then offers a significant improvement in tracking quality over output-based ADRC, even with the initial increased control

6.4 Error-Based ADRC

signal related to the time needed for the states to converge to proper values. It should be reminded that both ADRC schemes considered in this example do not explicitly utilize the information about the reference derivatives during the process of observer or controller synthesis, so this is not the case as seen earlier in Sect. 6.2.

The second potential reason for using the error-based variant of ADRC is, however, best described when the controller is represented in a transfer function form; hence we now move toward its input-output representation in s-domain.

> **Cooking Recipe (Error-Based ADRC, State-Space Form)**
>
> To implement and tune linear error-based ADRC in its continuous-time state-space form, the following steps have to be carried out:
>
> 1. *Plant modeling:* For the dominant dynamic plant behavior, identify the order N and the parameter b_0 of the plant model (3.2). Refer to Table 3.1 for some simple examples. Reformulate (3.2) into error-based form (6.22).
> 2a. *Controller tuning:* Error-based ADRC of order N requires tuning of N controller gains $k_{1,\ldots,N}$. When using *bandwidth parameterization* for the desired closed-loop dynamics, these gain values are obtained from (3.19) or Table A.1.1. If necessary, use an approximation as (3.21) to translate a time-domain specification into the tuning parameter ω_{CL}.
> 2b. *Observer tuning:* Error-based ADRC of order N requires $(N+1)$ observer gains $l_{1,\ldots,N+1}$. In case of *bandwidth parameterization*, choosing a relative observer bandwidth $k_{\text{ESO}} = 3, \ldots, 10$ is a good starting point. Use (3.24) or Table A.1.1 to compute the gains from k_{ESO} and ω_{CL}.
> 3. *Implementation:* Implement the state-space form of error-based ADRC as shown in Fig. 6.12. Required matrices and vectors are defined in the equations of the extended state observer (6.25) and the controller (6.27).

6.4.2 Transfer Function Representation

To derive the error-based ADRC variant in transfer function form, we start with the derivation of its closed-loop observer dynamics. To do that, the Laplace transform of the control law (6.27) is first put in the observer (6.26), which gives

$$\hat{\boldsymbol{x}}(s) = \left(s\boldsymbol{I} - \underbrace{\left(\boldsymbol{A} - \boldsymbol{l}\boldsymbol{c}^{\text{T}} - \frac{1}{b_0}\boldsymbol{b}\left(\boldsymbol{k}^{\text{T}}\ 1\right) \right)}_{\boldsymbol{A}_{\text{CL}}} \right)^{-1} \cdot \boldsymbol{l} \cdot e(s). \qquad (6.28)$$

To shorten further equations, we will use the abbreviated system matrix A_{CL} of the closed-loop controller/observer dynamics as introduced in (6.28):

$$A_{\text{CL}} = A - l c^{\text{T}} - \frac{1}{b_0} b \left(k^{\text{T}} \ 1 \right) = \begin{pmatrix} \mathbf{0}^{N \times 1} & \mathbf{I}^{N \times N} \\ 0 & \mathbf{0}^{1 \times N} \end{pmatrix} - \left(l \ \mathbf{0}^{(N+1) \times N} \right) - \begin{pmatrix} \mathbf{0}^{(N-1) \times N} & 0 \\ k^{\text{T}} & 1 \\ \mathbf{0}^{1 \times N} & 0 \end{pmatrix},$$

where A_{CL} has the same form as seen in (4.16). A transfer function of the controller can now be obtained by putting (6.28) back in the Laplace transform of (6.27), eliminating $\hat{x}(s)$ from the equations:

$$\begin{aligned} u(s) &= \frac{1}{b_0} \left(k^{\text{T}} \ 1 \right) \cdot (s\mathbf{I} - A_{\text{CL}})^{-1} \cdot l \cdot e(s) \\ &= \left(b_0 \cdot \det (s\mathbf{I} - A_{\text{CL}}) \right)^{-1} \cdot \left(k^{\text{T}} \ 1 \right) \cdot \operatorname{adj}(s\mathbf{I} - A_{\text{CL}}) \cdot l \cdot e(s). \end{aligned} \quad (6.29)$$

We can therefore put forth a control law for error-based ADRC using a single transfer function as follows, which is also visualized in Fig. 6.15:

$$u(s) = C_{\text{FB}}(s) \cdot e(s). \quad (6.30)$$

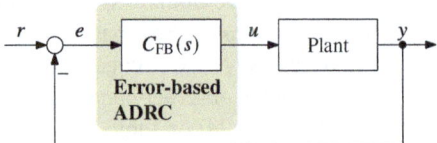

Fig. 6.15 Continuous-time error-based ADRC, transfer function form

In (6.30), term C_{FB} denotes the feedback controller transfer function, which can directly be read off (6.29):

$$C_{\text{FB}}(s) = \frac{\left(k^{\text{T}} \ 1 \right) \cdot \operatorname{adj}(s\mathbf{I} - A_{\text{CL}}) \cdot l}{b_0 \cdot \det(s\mathbf{I} - A_{\text{CL}})}, \quad (6.31)$$

which is a strictly proper transfer function that depends on the resolvent of A_{CL}.

Does the above derivation of the error-based ADRC in transfer function form look familiar? It should! We have already gone through a similar procedure for the "standard" output-based ADRC in Sect. 4.2. Through an independent derivation of the error-based variant, we have arrived at (6.31), which has exactly the same form as the feedback transfer function in (4.20) for the output-based case. This means that we can conveniently reuse here the structures and parameters already defined for output-based ADRC. Consequently, the feedback controller transfer function (6.31) of error-based ADRC can be alternatively written as

6.4 Error-Based ADRC

$$C_{FB}(s) = \frac{K_I}{s} \cdot \frac{1 + \sum_{i=1}^{N} \beta_i \cdot s^i}{1 + \sum_{i=1}^{N} \alpha_i \cdot s^i},$$

which is identical to (4.22) with exactly the same coefficients α, β, and gain K_I, all of which were already introduced and are available from Appendix A.2.

The above similarities reveal there are strong connections between the output- and error-based ADRC structures. By exploring those connections, let us identify potential benefits error-based ADRC can bring. The transfer function implementation of error-based ADRC is depicted in Fig. 6.15. By comparing it with the previously introduced transfer function representations of control loops with continuous-time ADRC in Figs. 4.2 and 4.3, one can notice that in the error-based case, due to the injection of feedback error as the algorithm input signal, the control law is trivialized to just C_{FB}. Since the reference signal is not directly related to the derivation of the error-based ADRC algorithm, there is no need for the prefilter (meaning $C_{PF} = 1$), which results in the control signal simply being (6.30). Therefore, the important practical issue of transfer function realizability, previously considered in Sect. 4.2, is nonexistent here, as the sole transfer function C_{FB} in (6.30) is always realizable.

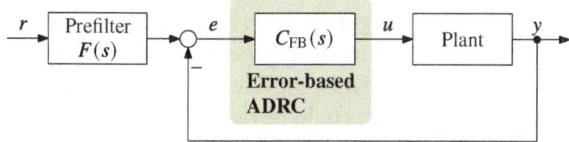

Fig. 6.16 Continuous-time error-based ADRC, transfer function implementation, with a custom prefilter $F(s)$ to decouple the design of reference tracking and disturbance rejection

However, the most important reason why the error-based variant of ADRC has been introduced in this book is that it gives more freedom to the control designer to construct custom control solutions, since the prefilter C_{PF} is not built-in (imposed) in the design in advance—as it is the case in "standard" output-based ADRC. If needed, one can prepare a custom prefilter and add it to the reference channel of error-based ADRC to gain the possibility of shaping certain desired transient system behavior. Customizing error-based ADRC with a user-defined prefilter, shown in Fig. 6.16 and denoted as $F(s)$, is recommended for true 2DOF possibilities. To visualize that, we have prepared an example, the results of which are seen in Fig. 6.17, where one can notice the influence of the user-defined prefilter on the system output.

Regarding the connection between error-based and output-based ADRC schemes, the presence of C_{PF} in output-based ADRC has no effect on the stability margins, sensitivity to measurement noise, or disturbance rejection but has an impact on the steady-state reference tracking performance as shown in Fig. 6.17. In short, it can be seen, for example, from Fig. 6.15, that the error-based ADRC has a simpler, more

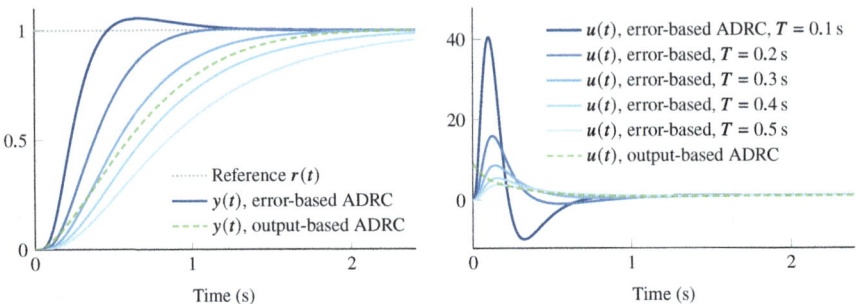

Fig. 6.17 Closed-loop step responses of output-based ADRC (dashed lines) and error-based ADRC (solid lines), the latter combined with reference signal filters $F(s) = \frac{1}{T^2 s^2 + 2T s + 1}$ ranging from $T = 0.1$ s (darker colors) to $T = 0.5$ s (brighter colors). The plant has a normalized second-order low-pass behavior $P(s) = \frac{1}{s^2 + 2s + 1}$. Tuning parameters are $N = 2$, $\omega_{CL} = 3$ rad/s, and $k_{ESO} = 10$. Using customized prefilters for error-based ADRC enables true 2DOF design possibilities, whereas the output-based ADRC design is fixed at a critically damped response

compact design, resulting in a simpler implementation (due to the lack of prefilter), but for the price of reducing the control over transient tracking performance.

Before we move on to the next topic, there is also an interesting characteristic of the error-based ADRC that we did not mention before. The reformulation of ADRC in error-based form allows for its direct comparison with existing industrial controllers, like PI and PID, which are also often expressed in error-based form. This is potentially helpful to those who are interested in fostering the adoption of ADRC in industrial practice as a viable alternative to standard controllers. Expressing ADRC in a form, which is familiar to the field engineers, could be one of its selling points.

> **Cooking Recipe (Error-Based ADRC, Transfer Function Form)**
>
> To implement and tune the transfer function form of error-based ADRC derived in this section, the following steps have to be carried out:
>
> 1. *Plant modeling:* Identify the order N and the gain parameter b_0 of the plant model.
> 2. *Controller and observer tuning:* When using *bandwidth parameterization*, the tuning parameters are ω_{CL} and k_{ESO}. For first- and second-order error-based ADRC, these values can simply be put in the design equations for the K_I and α, β coefficients tabulated in Tables A.2.2 and A.2.3 on page 194, respectively. It is also possible to compute these coefficients from the otherwise obtained state-space controller and observer gains $\boldsymbol{k}^T$ and $\boldsymbol{l}$.
> 3. *Implementation:* The structure of the control law (6.30) is shown in Fig. 6.15, consisting solely of a feedback controller $C_{FB}(s)$. If needed, a custom prefilter $F(s)$ could be optionally attached in the reference
>
> (continued)

6.4 Error-Based ADRC

signal path, as shown in Fig. 6.16, to help with the transient tracking performance. The feedback controller $C_{\text{FB}}(s)$ contains an integrator, which, to avoid windup issues, should be implemented with output limits.

Open Access This chapter is licensed under the terms of the Creative Commons Attribution 4.0 International License (http://creativecommons.org/licenses/by/4.0/), which permits use, sharing, adaptation, distribution and reproduction in any medium or format, as long as you give appropriate credit to the original author(s) and the source, provide a link to the Creative Commons license and indicate if changes were made.

The images or other third party material in this chapter are included in the chapter's Creative Commons license, unless indicated otherwise in a credit line to the material. If material is not included in the chapter's Creative Commons license and your intended use is not permitted by statutory regulation or exceeds the permitted use, you will need to obtain permission directly from the copyright holder.

Chapter 7
Interlude: A Look Around

Abstract Equipped with an essential understanding of ADRC, one could confidently move to the second part of the book, which deals with its practical implementation. But for those wishing to get a wider look at the topic of ADRC and get a bit more context than what has been provided in the book so far, we use this chapter to take a quick pause and *look around*. Here we recall what we covered so far in the first part, put things in historical perspective, and provide relevant bibliographical support. We also briefly touch on some of the topics in the area of ADRC, which, although interesting and horizon-broadening, go beyond the scope of this book, which is focused on fundamentals. Finally, we discuss what is to come in Part II.

A Look Back

Part of looking around is looking back. By reexamining the first six chapters of the book, one could notice that no literature references were provided alongside the presented information on ADRC. This was intentional on our end because we wanted to provide an introduction to ADRC without recurrent interruptions caused by referring to external materials. Now, as we are about to finish the first part of the book dedicated to the theoretical foundations of ADRC, it feels like a good time to review the so far presented material and give credit where credit is due. We named this entire chapter "Interlude" for a reason as we want to give the reader a bit of a break from the technical regime of the past few chapters and time to reflect on the things presented thus far. Here we want to give a literature overview and a historical perspective on ADRC for things we have shown in the first part. Usually placed at the beginning of a book, here such an overview is deliberately delayed as only now we have introduced enough information on ADRC for such a rundown to be meaningful to the readers. Therefore, let us now go through the chapters one by one. Before proceeding, we stress once again that even though we will be giving references to external materials, this book is written to be a self-contained introduction to ADRC, and you do not have to read anything else to get the gist of ADRC.

- **Chapter 2 (First Contact with ADRC)**: In retrospect, the two illustrative examples from Chap. 2 were treated with a similar state-space variant of ADRC, consisting of a linear Luenberger-like observer and a linear state-feedback controller. In fact, the idea of utilizing linear ADRC comes from Prof. Zhiqiang Gao's seminal work [1]. It keeps the ADRC design and implementation process simple. Thanks to the relatively high level of robustness of even the linear ADRC structure, it is a good starting point for finding a candidate solution and also for more challenging control problems. The justification for that can be found in engineering practice, which favors simplicity, assuming the control system performs with acceptable quality. For the same reason, we focus so much in this book on the linear ADRC variant, which balances straightforwardness with effectiveness.
- **Chapter 3 (Linear Active Disturbance Rejection Control)**: Here we generalized the first- and second-order examples from the previous chapter and introduced more mathematical rigor and control domain-related terminology in our explanation. Intuitively, the roots of the ADRC form derived in this chapter can also be traced back to the breakthrough paper [1], and it is safe to say that this chapter was built on that key work. In the general linear ADRC form, presented in Chap. 3, we advocated for the use of so-called *bandwidth parameterization* for tuning both the observer and the controller parts (see Sect. 3.2), which is based on a conventional pole placement approach. This idea, also coming from [1], not only facilitates stability analysis but also significantly reduces the number of tuning parameters as all controller and observer gains can be parameterized as functions of the control and observer bandwidths. Thanks to its benefits, the *bandwidth parameterization* is now considered the predominant tuning approach in linear ADRC. The rule of thumb formulas, like the ones from [2] and shown in Sect. 3.2.1, relating the settling time and bandwidth, help to further streamline the design and implementation of linear ADRC (which will become apparent later in the book). Regarding the notation introduced in Sect. 3.2.2, the term k_{ESO} also comes from [2] and is the relative observer bandwidth factor, which helps to express the observer bandwidth as a dependent of the closed-loop bandwidth. Also in this chapter, term b_0 is referred to as the *critical gain parameter*, a well-deserved name, received in [3], based on its importance on the overall control performance, which in this book is later visualized and made clear in Chap. 5.
- **Chapter 4 (Between Time and Frequency Domains)**: First of all, the name of the chapter was inspired by Zheng and Gao [4], where the authors discussed selected results in the analysis of linear ADRC and offered interpretations in frequency domain, which tends to be more familiar to practicing engineers. In this chapter, we were on a similar path. To this end, as done in [2], we wanted to precisely reveal the connection of linear ADRC to "classical" state-space control. For completeness, an important note has to be added to the discussion of the elements of controller (4.5). In Sect. 4.1, we claimed that inserting (4.5) in (4.1) reveals that the disturbance may be fully rejected if accurately estimated and if $\boldsymbol{b} \cdot \boldsymbol{k}_{\text{d}}^{\text{T}} = \boldsymbol{e} \cdot \boldsymbol{c}_{\text{d}}^{\text{T}}$ can be achieved. This claim, however, is supported by rigorous proof that can be found in [5]. From the same work comes the generic structure of an observer-based state-feedback control loop with scalar input/output signals, which

we used to draw Fig. 4.1. When discussing the form of the feedback controller $C_{FB}(s)$ in Sect. 4.2.1, it was important to us to highlight (as we strive to make the book close to practice) that for orders $N = 1$ and $N = 2$, it is structurally equivalent to the so-called Type 2 or Type 3 controllers used in power electronics [6]. These are, in turn, equivalent to PI or PID controllers equipped with an additional first- or second-order low-pass filter, as recommended by Hägglund [7]. When it comes to specific mathematical operations performed in this chapter, like the alternative derivation of $C_{FF}(s)$ and $C_{PF}(s)$ from (4.18), shown in Sect. 4.2.2, they were adopted from an already existing work [8]. Finally, in this chapter, there was talk about "paradigm shift" in the context of ADRC. This expression was used deliberately as it is a direct callback to work [9], which argues that the departure from the established model-based control school, offered by ADRC, is a "paradigm shift" in the area of control and ADRC's key selling point.

- **Chapter 5 (Visual Tour)**: In this chapter, we wanted to display the capabilities and limitations of the ADRC scheme. This included the ability to follow reference signals and compensate for the effects of load disturbances and process variations and to check the influence of measurement noise. We decided that these properties will be captured by the *gang of six* transfer functions, an analysis tool that was made popular through [10]. From the same work, we know that the RHP zeros severely restrict the achievable bandwidth of a control system, and the effects seen in Fig. 5.17 were no exception to that. It is worth mentioning that the idea of an in-depth analysis of the first-order ADRC, surrounding (5.8) and involving putting the coefficients for *bandwidth parameterization* from Table A.2.2 in $C_{FB,1}(s)$, was taken from [8]. Finally, if interested in details regarding the footnote on page 70 commenting on the control design and the very typical range for filter damping, the reader could visit [11].
- **Chapter 6 (Extensions and Modifications)**: Here we explored what else you can do with ADRC that goes beyond what was presented in the first few chapters (and therefore implicitly also in [1]). From the vast literature on ADRC, we have selected a few design choices that, to us, offer interesting improvements to the base linear ADRC and may become handy for future control practitioners. The first one was the availability of additional model information, covered in Sect. 6.1. It is quite intuitive that one can take advantage of the known plant information to improve the performance of a conventional ADRC since the more you know about the plant, the "smaller" the *total disturbance* is. This relation was systematically studied in [12], where the extra information about the controlled system model (when available) was shown to improve the ADRC performance, especially for unstable, time-delayed, and non-minimum phase processes. The second option was related to the use of reference signal derivatives in the ADRC design process, discussed in Sect. 6.2. While the incorporation of the reference signal derivatives in the ADRC design process is quite straightforward, the question of how to get those signals is a rather challenging one as they may not be easily available in practice. Signal derivatives reconstruction is a research subject on its own, and an overview of some of the available techniques, differing in complexity and effectiveness, can be found in [13]. Then, in Sect. 6.3, we have shown a potential

improvement to the core linear ADRC based on adding nonlinear components, which improve the convergence rate for the price of increased implementation complexity and extra tuning efforts. The version of nonlinear ADRC shown in that section, consisting of a nonlinear TD and a nonlinear weighting function, was the one from [14]. Interestingly, the presented TD is a signal differentiator hence could also be considered as a potential candidate to estimate higher-order reference derivatives, mentioned above. Instinctively, the nonlinear elements presented in this section are just examples. Literature is now rich with different nonlinear tools that could be used to improve the performance of ADRC, that is, of course, if the base linear ADRC does not provide satisfactory results. An overview of different nonlinear add-ons can be found in [15]. Finally, we decided to include in the book an alternative, error-based version of ADRC, discussed in Sect. 6.4. Although the error-based formulation of ADRC was used for theoretical analysis [16], facilitated by the entire control system being conveniently expressed in feedback error dynamics form, later the practical benefits of having ADRC in error-based form started to be explored as well [17, 18]. This led to many success stories of the application of error-based ADRC in practice in fields like robotics [19] and power electronics [20, 21]. More about the connection between error-based and output-based ADRC can be found in [22].

The phrase "look back" is understood here twofold. It is not only looking back at what was covered so far in the book by putting things in perspective and providing relevant bibliographical support but also looking back at the history of ADRC. To this end, we have been mostly focused on a specific variant of ADRC, one that uses a linear extended state observer and a linear state-space controller, in other words, a variant that we have been referring to as "LADRC." However, it should be noted that the area of ADRC did not start with a linear version.

Tracking back the roots of ADRC, one can find that, as a concept, it was initially introduced by Prof. Jinqing Han as a nonlinear scheme [14], similar to the one we discussed in Sect. 6.3.[1] It was only later streamlined through the linear variant [1]. To truly grasp the entirety of the long way ADRC has come, we recommend overview papers [23, 24], where the making of ADRC is described in detail. They offer a historical (and philosophical at times) perspective on the origins of ADRC, revisit how this paradigm came into being, and finally reexamine what helped ADRC to make the transition from an idea to an industrial technology. We intentionally leave to those articles the comprehensive telling of the story of ADRC until the point we pick it up. So when does our book enter this story which spans decades? And the answer is clear: 2003. This is the year when LADRC, with its convenient *bandwidth parametrization*-based tuning, was introduced in [1]. This is the point in time when we arrive and pick things up in the book. And finally, this is the variant of ADRC

---

[1] The ADRC methodology has been forming since the 1990s with initial results documented in Chinese. Reference [14] from 2009 is often cited as the introduction of ADRC to the mainstream English-speaking audience. To better understand the origins of ADRC and how this idea was forged and evolved, the readers are referred to the pioneering work [14] and the references therein.

that we have been almost exclusively focused on so far in the book. But why this one? Are not there more powerful ADRC forms out there already?

One of the reasons behind such a choice is that the linear version allows a relatively simple introduction to the entire topic of ADRC and helps to form a solid base on which one can then build more advanced and more customized solutions. Our aim in writing this book was to put forward a text that develops along a single line of argument, hence the idea of using LADRC as a gateway to the world of ADRC seemed natural. This focused take helped us to tell a coherent story across Chaps. 2–6 and not worrying about distracting the reader with different variants or going off on a tangent with what else one can do with ADRC. It also allowed the reader to stay focused and understand the core concepts of the introduced methodology. Our objective is not to explore the full depth of mathematical completeness or cover all the nuances of ADRC but instead to give enough detail so that a reader can begin applying this approach as soon as possible. Focusing on a simple, fundamental form of ADRC thus helps to provide a sufficiently strong foundation, so that the reader can comfortably turn to the study of appropriate complementary literature on ADRC, and the LADRC variant facilitates that. The other reason behind focusing mostly on linear state-space ADRC in Part I, besides being straightforward, was shown in many documented instances to provide satisfactory results even for some complex control problems. Engineering practice is dominated by the linear variant. It is also from our own experience as we have been putting things to work with ADRC for years now. The undemanding structure and the limited number of tuning gains in LADRC paved the way for its widespread adoption. Therefore this variant fits well with the introductory style of this book.

A Look Beyond

It should be clear by now that, for reasons explained above, we focus in the book on a particular form of ADRC, which results from certain design choices we have made at the beginning, for example, in terms of choosing the plant model, disturbance model, observer structure, type of controller, and tuning approach. Now, we would like to take the opportunity to share some personal recommendations for further reading for those who wish to study ADRC in different aspects that go beyond this book. With this, we aim to broaden the readers' horizons by showing what else the world of ADRC has to offer. Since the area of ADRC has been exponentially growing over the years, its current body of knowledge is massive. We had to thus apply certain criteria for the selection of materials we would like to recommend. In general, we focused on what else could you do in ADRC to arrive at roughly the same point as in this book. As a result, the below list provides literature support for things not covered in the book but potentially interesting after its completion. Furthermore, their selection is bound to involve compromise as we have not attempted to review all the results on ADRC that could be construed as being relevant.

Different Observer Types: Until this point in the book, we have mostly focused on the simple, linear, Luenberger-like ESO (except for the nonlinear variant mentioned in Sect. 6.3). But other observer structures may be used as well, each having its own set of advantages and disadvantages, as discussed in overview papers [25–27]. For example, if one changes the *total disturbance* model in ADRC, then the observer design would naturally have to be aligned. Therefore, one can find in the literature extended linear forms such as *generalized proportional integral* observers, utilizing polynomial models for the *total disturbance* [28] or resonant ESOs, utilizing harmonic models [29–31]. If one is interested in addressing specific aspects of the observer operation, there are specific structures that can accommodate, for example, those modifying the convergence rate by utilizing sliding modes [32], which in some cases can also offer finite-time convergence [33], or variants that can minimize the observer peaking phenomenon [34]. Furthermore, if certain signals related to the plant are available and can be utilized in the process of observer synthesis, then one could go for a reduced-order ESO to lower the overall complexity of the control system [35, 36]. There is also the idea of combining several observers within one ADRC design but is reserved for specific systems that can handle such increased computational burden. The multiobserver designs are commonly realized by connecting the observers in a cascade [37–41] for extra denoising.

Different Controller Types: In the book, we have limited ourselves to the use of simple, classical controllers, like the ones seen in Chap. 3. Similar to the observer case, the reason behind such a choice was to keep the design and implementation simple and to stick with P and PD controllers, which can be found in any textbook on control. But in general, choosing the controller type in ADRC is up to the user. It is, however, a good engineering practice to introduce more advanced (and potentially more complicated) tools only if there is a clear practical need and justification for that. In [28], for example, the authors exploited the plant's flatness property and designed high-order controllers called *generalized proportional integral* controllers. Another example can be found in [42], where the controller was designed using the sliding mode theory. There are many options to choose from related to how the controller is placed in the control system (example being a cascade control in [43]), what is the algorithm behind the calculation of the control signal (example being MPC in [41]), and to what specific system the ADRC is applied to (example being a controller tailored to high-order integral systems in [44]).

Different Tuning Methods: Without a doubt, the *bandwidth parameterization* tuning methodology has streamlined the development of ADRC. Considering its ease of use and reported satisfactory results in many cases, it is not surprising that it is the most common choice. We, therefore, rely on it throughout this book. But we also want to mention some of the options for choosing observer gains. Even though relatively little literature is available on alternative tuning methods, one can find some customized solutions, like *characteristic ratio assignment* to directly control the transient response [45] or those dedicated to time-delayed systems [46, 47]. One can also find in the literature some tuning approaches dedicated to very specific control problems that call for, for the price of increased computational complexity, more advanced tools like neural networks [48] or genetic algorithms [49].

Theoretical Results: Since we decided in the book to focus on the simple linear variant of ADRC, we can utilize classical linear systems theory, including pole analysis, to determine the stability of the control system. Those simple tools from the linear framework, taught in every control systems introductory course, apply to everything we have shown so far regarding LADRC. However, complex analysis, involving strong mathematical tools, is needed to verify more complex ADRC variants. The advanced theoretical results in ADRC mostly consider the study of observer convergence, stability of the closed-loop system, and robustness analysis (against parametric uncertainties, external disturbances, and unmodeled dynamics of the system) with papers [16, 50, 51] giving an overview of the current theoretical scene. One of the still challenging topics in the area of ADRC is establishing the applicability conditions for various classes of plants, with some progress made utilizing the tools of differential geometry [52, 53]. One of the key aspects of any ADRC design is the formulation of the *total disturbance* term, and works like [54] provide valuable insight into it.

Similar Control Methodologies: The idea of online estimation and rejection of *total disturbance* as opposed to relying on the necessity of having a detailed mathematical model of the controlled plant is not unique to ADRC. One can find other methodologies that handle control problems in such a way. Overview papers, like [55, 56], present the landscape of available approaches and the history of their evolution. The specific techniques mostly differ in the way how the aggregated uncertainty is being reconstructed. While in ADRC we use state observer for *total disturbance* estimation, in approaches like *model-free control*,[2] it is obtained with the means of algebraic estimation [57], in *disturbance observers-based control* schemes using properly constructed low-pass filters [58], or in *balance-based adaptive controller* using recursive least-squares procedure [59].

Equivalence with Other Control Methods: After the ADRC started to gain traction among the control community, a natural discussion started on how to position ADRC in the control landscape. Such discussion led to the research on finding equivalences with already existing control schemes. To this end, theoretical connections were found between ADRC and standard industrial controllers, like PI [60] and PID [61, 62] and also with more advanced control structures [63, 64]. Multiple researchers have shown independently that, with some simplification and for low-order plants, ADRC is indeed backward compatible with PI/PID. There is also now an interesting body of work trying to derive ADRC tuning rules from the PID parameters [65] or attempting to interpret classic controllers as disturbance observer-based structures [66]. A connection between error-based ADRC and classic industrial controllers was studied in [67]. Finding backward compatibility of ADRC, especially with PI and PID, is an important problem to solve as the ignorance of such an impediment arguably led to the stagnation of many of the advanced controllers and to the questioning of their relevance to engineering practice [68].

---

[2] The name means that the design process is "free" from accurate knowledge about the plant. The same misconception sometimes surrounds ADRC, which is falsely called a model-free approach, even though some rough plant information is obviously needed, like the order of the plant, to proceed with controller and observer syntheses.

Other Books on ADRC: There are several books on ADRC on the market at the moment. They differ from each other in terms of focus and target audience. To enable a comparison with our work, we therefore want to give a brief overview. Our book is a concise, from scratch, step-by-step introduction to the world of ADRC where things are comprehensively explained from basic theory all the way to practical implementation. It stands on its own as a complete work. Hence, we view the following books as beneficial for those wishing to go beyond our book and extend their horizons by learning more about specific aspects of ADRC. We discuss them in the order of their publication year. In 2008, the first book on ADRC appeared and was published by the already mentioned Prof. Han [69]. In the book, written in Chinese, Han meticulously discussed all aspects of the original, nonlinear ADRC and its components. The book embraces the power of nonlinear feedback and puts it to full use through its incorporation in observer, controller, and tracking differentiator. The nonlinear ADRC is proposed there as a combination of both worlds, classical feedback control and modern control theory, which formulates a direct response to the limitations of PID. In 2014, a book was published, which offered a wider look at the topic of disturbance estimation and rejection [70]. The authors combined their extensive theoretical and practical knowledge and gave an overview of various techniques (not only ADRC) that can be used when the standard disturbance compensation via feedforward is not an option. The book not only contains the main concepts and design philosophies of various disturbance observer-based control approaches but also shows multiple simulation and experimental results. In 2016, a book was published that offered strong mathematical support for ADRC, including high-level convergence and stability proofs [71]. The focus of the book was on nonlinear systems and because of that covered a variety of advanced topics. The included theoretical derivations help to highlight the advantages of ADRC, including small overshooting, fast convergence, and energy savings. In 2017, two important books on ADRC were introduced to the market. The main focus of [72] was on exploring the relation between the development of information obtaining and processing technologies and the ability to merge various types of disturbances into one "equivalent" disturbance. The book also offers a summary of various developments in anti-disturbance control and focuses on descriptions of various control and filtering strategies. The second book [28] discusses ADRC as a methodology that differs from current robust feedback controllers, characterized by complex matrix manipulations, complex parameter adaptation schemes, and in other cases, induced high-frequency noises through the classical chattering phenomenon. The book does the above dissection using the notion of differential flatness. In 2021, a book on ADRC was published which had a channeled focus on a specific type of ADRC [73]. It investigated the theoretical efficiency and performance of so-called event-triggered ADRC, i.e., for nonuniform sampled-data control.

Software Support for ADRC Implementation: Since almost every commercial control equipment provides nowadays hardware or software PID function block supporting its fast and reliable implementation, some research was inevitably dedicated to facilitating the deployment of ADRC in industrial settings as well. In [74], the PLC-based implementation of ADRC was presented in the form of a general-purpose

7 Interlude: A Look Around   111

function block, allowing to choose the order of ADRC, switch between ESO and
generalized proportional integral observer (with some user-defined order), and also
switch between linear and nonlinear observer type (by properly setting the $f_{al}(\cdot)$
function parameters). In [75], a similar idea of a flexible function block for PLC was
presented and offered some more practically oriented add-ons like derivative backoff
and a built-in tuning approach based on the process step response.

Discrete-Time ADRC: A practical implementation of ADRC will almost always
be in the form of software. Therefore, given the inherent discrete-time nature of the
underlying target processor systems, it is necessary to provide discrete-time variants
of ADRC for actual deployment in hardware. At the moment, readers interested in
the use of discrete-time ADRC usually come across paper [76], in which the authors
investigated and compared various digital implementations of the ESO. Over the
years, there were also some customized ADRC solutions in discrete domain. For example, in the already mentioned book [73], the application of discrete-time ADRC
was considered for the special case of event-triggered systems. In [77], the authors
discretized the ESO using a discrete delta operator, which created a rapprochement
between continuous and discrete dynamic system models and established a natural
framework to investigate the behavior of discrete dynamic models in fast sampling
limit. Surprisingly, the literature on discrete ADRC for uniform sampled-data control is quite scarce, given its crucial importance in practical implementation and
deployment. This motivated us to work on this important subject and fill this gap.
This is why we dedicate the entire upcoming second part of the book to this topic
where we also present our results in that area.

Commercial Interest: The practically appealing high effectiveness-to-complexity
ratio of ADRC has inevitably led to the first signs of increased interest in it from
the industry. For example, Texas Instruments, a global semiconductor company, has
incorporated the ADRC approach in some of its motion control products.[3] Another
example is the inclusion (since release R2022b) of an official ADRC library block
in MATLAB. We will take a deeper look into this block in Appendix B.

Looking Forward

Truth be told, all that we have shown so far in the book could be learned from
scientific papers and books. We did, however, put our spin on things by offering our
unique perspective (combining years of experience in academia and industry), structure, and way of presentation. But part of looking around is also looking forward.
To this end, Part I, which dealt with fundamentals and continuous-time domain, is
done, and now we are going practical with ADRC. And putting this into practice, as
mentioned above, is really something different and can basically be done nowadays
only in discrete time. Looking at the available literature, including the last couple of
bullet points in the above further readings list, going into actual discrete implemen-

---

[3] An example of using ADRC by Texas Instruments is the InstaSPIN technology dedicated to
brushless DC motors (https://www.ti.com/tool/INSTASPIN-BLDC).

tation is not often done in terms of ADRC, and there are many open points when it comes to ironing out edges to facilitate practice-oriented implementation. Although there are many works documenting successful implementations of ADRC in practice, the literature investigating the foundations of ADRC in discrete domain seems underdeveloped. In our subjective estimation, more than 90% of ADRC-related papers either deal with its continuous-time and its original nonlinear variant or present only simulation results. There are a lot of materials on ADRC, some of which we collected here, but the air is very thin when it comes to discrete implementation and, as we saw earlier, the topic is undercovered in the ADRC landscape. This is where we see our main working field, both in our past works and in this book, as we wanted to contribute something in that area. This is also why the selling point of this book lies in its dual nature, in which necessary basics are first introduced in Part I, and only then practical implementation is covered in Part II, as it needs a separate, dedicated treatment. Concrete solutions for the application-oriented digital applications of ADRC in the discrete form are exactly what is yet to come in the book. We will need to, however, introduce a few more references in the second part of the book, but the style of reference will be different than in Part I as we will provide citations in the text whenever it is needed, not delaying it this time to the end of Part II. The second difference is that most of the cited references in Part II are our own works because we have been in this space and this has been our research interest for years.

Therefore, the second part of the book will be primarily based on our works [8, 78–80]. Let us quickly break them down. In [8], to enable immediate comparability with existing classical control solutions and to support the adoption of ADRC in industrial practice, we have introduced a realizable transfer function implementation of continuous-time linear ADRC. In the same work, an exact implementation of discrete-time ADRC using transfer functions was introduced for the first time, with special emphasis on practical aspects such as computational efficiency, low parameter footprint, and windup protection. Then, in [78], we introduced an even more efficient implementation than the transfer function, which focused on minimizing the number of storage variables and algebraic operations. This "minimum-footprint implementation" remains the most efficient implementation to date. In [79], we have provided various options for discrete implementations of error-based ADRC, even with the same set of parameters as in the output-based variants. Finally, in [80], the proposed *half-gain tuning* results in similar closed-loop dynamics as the commonly employed *bandwidth parameterization* design, but with lower feedback gains, paving the way for using ADRC in more noise-affected applications. Although originally presented in continuous time, we will use it in Part II in its discrete form.

Before you transition to the second part of the book, be reminded once again that you do not need any of the above-cited works on discrete ADRC moving forward as we will provide all the necessary details. We want to keep the book self-contained also in the second part, so you can move through it without having to read any of the references mentioned in the upcoming chapters.

References

1. Gao, Z.: Scaling and bandwidth-parameterization based controller tuning. In: Proceedings of the American Control Conference, pp. 4989–4996 (2003). https://doi.org/10.1109/ACC.2003.1242516
2. Herbst, G.: A simulative study on active disturbance rejection control (ADRC) as a control tool for practitioners. Electronics **2**(3), 246–279 (2013). https://doi.org/10.3390/electronics2030246
3. Madonski, R., Gao, Z., Łakomy, K.: Towards a turnkey solution of industrial control under the active disturbance rejection paradigm. In: Proceedings of the Annual Conference of the Society of Instrument and Control Engineers of Japan, pp. 616–621 (2015). https://doi.org/10.1109/SICE.2015.7285478
4. Zheng, Q., Gao, Z.: Active disturbance rejection control: between the formulation in time and the understanding in frequency. Control Theory Technol. **14**(3), 250–259 (2016). https://doi.org/10.1007/s11768-016-6059-9
5. Ostertag, E.: Mono- and Multivariable Control and Estimation. Springer, Berlin (2011). https://doi.org/10.1007/978-3-642-13734-1
6. Venable, H.D.: The K factor: a new mathematical tool for stability analysis and synthesis. In: Proceedings of the National Solid-State Power Conversion Conference (1983)
7. Hägglund, T.: Signal filtering in PID control. IFAC Proceedings Volumes **45**(3), 1–10 (2012). https://doi.org/10.3182/20120328-3-IT-3014.00002
8. Herbst, G.: Transfer function analysis and implementation of active disturbance rejection control. Control Theory Technol. **19**, 19–34 (2021). https://doi.org/10.1007/s11768-021-00031-5
9. Gao, Z.: Active disturbance rejection control: a paradigm shift in feedback control system design. In: Proceedings of the American Control Conference, pp. 2399–2405 (2006). https://doi.org/10.1109/ACC.2006.1656579
10. Åström, K.J., Murray, R.M.: Feedback Systems: An Introduction for Scientists and Engineers, 2nd edn. Princeton University, Princeton (2021)
11. Larsson, P.O., Hägglund, T.: Control signal constraints and filter order selection for PI and PID controllers. In: Proceedings of the American Control Conference, pp. 4994–4999 (2011). https://doi.org/10.1109/ACC.2011.5991112
12. Fu, C., Tan, W.: Tuning of linear ADRC with known plant information. ISA Trans. **65**, 384–393 (2016). https://doi.org/10.1016/j.isatra.2016.06.016
13. Zhang, H., Xiao, G., Xie, Y., Guo, W., Zhai, C.: Tracking Differentiator Algorithms: Theories, Implementations and Applications. Lecture Notes in Electrical Engineering. Springer, Singapore (2021). https://doi.org/10.1007/978-981-15-9384-0
14. Han, J.: From PID to active disturbance rejection control. IEEE Trans. Ind. Electron. **56**(3), 900–906 (2009). https://doi.org/10.1109/TIE.2008.2011621
15. Madonski, R., Herman, P.: Survey on methods of increasing the efficiency of extended state disturbance observers. ISA Trans. **56**, 18–27 (2015). https://doi.org/10.1016/j.isatra.2014.11.008
16. Huang, Y., Xue, W.: Active disturbance rejection control: methodology and theoretical analysis. ISA Trans. **53**(4), 963–976 (2014). https://doi.org/10.1016/j.isatra.2014.03.003
17. Michałek, M.M.: Robust trajectory following without availability of the reference time-derivatives in the control scheme with active disturbance rejection. In: Proceedings of the American Control Conference, pp. 1536–1541 (2016). https://doi.org/10.1109/ACC.2016.7525134
18. Madonski, R., Shao, S., Zhang, H., Gao, Z., Yang, J., Li, S.: General error-based active disturbance rejection control for swift industrial implementations. Control. Eng. Pract. **84**, 218–229 (2019). https://doi.org/10.1016/j.conengprac.2018.11.021
19. Chen, S., Chen, Z., Zhao, Z.: An error-based active disturbance rejection control with memory structure. Meas. Control **54**(5-6), 724–736 (2021). https://doi.org/10.1177/0020294020915219

20. Madonski, R., Łakomy, K., Yang, J.: Simplifying ADRC design with error-based framework: case study of a DC-DC buck power converter. Control Theory Technol. **19**, 94–112 (2022). https://doi.org/10.1007/s11768-021-00035-1
21. Huang, C., Zhao, H.: Error-based active disturbance rejection control for wind turbine output power regulation. IEEE Trans. Sustainable Energy **14**(3), 1692–1701 (2023). https://doi.org/10.1109/TSTE.2023.3243386
22. Madonski, R., Herbst, G., Stankovic, M.: ADRC in output and error form: connection, equivalence, performance. Control Theory Technol. **21**, 56–71 (2023). https://doi.org/10.1007/s11768-023-00129-y
23. Gao, Z.: On the centrality of disturbance rejection in automatic control. ISA Trans. **53**(4), 850–857 (2014). https://doi.org/10.1016/j.isatra.2013.09.012
24. Gao, Z.: Active disturbance rejection control: from an enduring idea to an emerging technology. In: Proceedings of the International Workshop on Robot Motion and Control, pp. 269–282 (2015). https://doi.org/10.1109/RoMoCo.2015.7219747
25. Radke, A., Gao, Z.: A survey of state and disturbance observers for practitioners. In: Proceedings of the American Control Conference (2006). https://doi.org/10.1109/ACC.2006.1657545
26. Łakomy, K., Patelski, R., Pazderski, D.: ESO architectures in the trajectory tracking ADR controller for a mechanical system: a comparison. In: Advanced, Contemporary Control, pp. 1323–1335. Springer, Berlin (2020). https://doi.org/10.1007/978-3-030-50936-1_110
27. Khalil, H.K.: Extended high-gain observers as disturbance estimators. SICE J. Control Meas. Syst. Integr. **10**(3), 125–134 (2017). https://doi.org/10.9746/jcmsi.10.125
28. Sira-Ramírez, H., Luviano-Juárez, A., Ramírez-Neria, M., Zurita-Bustamante, E.W.: Active Disturbance Rejection Control of Dynamic Systems: A Flatness Based Approach. Butterworth-Heinemann, Oxford (2017). https://doi.org/10.1016/C2016-0-01983-6
29. Stankovic, M., Madonski, R., Shao, S., Mikluc, D.: On dealing with harmonic uncertainties in the class of active disturbance rejection controllers. Int. J. Control. **94**(10), 2795–2810 (2021). https://doi.org/10.1080/00207179.2020.1736639
30. Madonski, R., Łakomy, K., Stankovic, M., Shao, S., Yang, J., Li, S.: Robust converter-fed motor control based on active rejection of multiple disturbances. Control. Eng. Pract. **107**, 104696 (2021). https://doi.org/10.1016/j.conengprac.2020.104696
31. Li, J., Sun, M., Chen, Z.: An add-on damping enhancement with adjustable gain for lightly damped system. Mech. Syst. Signal Process. **191**, 110179 (2023). https://doi.org/10.1016/j.ymssp.2023.110179
32. Dai, C., Yang, J., Wang, Z., Li, S.: Universal active disturbance rejection control for nonlinear systems with multiple disturbances via a high-order sliding mode observer. IET Control Theory Appl. **11**(8), 1194–1204 (2017). https://doi.org/10.1049/iet-cta.2016.0709
33. Han, L., Mao, J., Cao, P., Gan, Y., Li, S.: Toward sensorless interaction force estimation for industrial robots using high-order finite-time observers. IEEE Trans. Ind. Electron. **69**(7), 7275–7284 (2022). https://doi.org/10.1109/TIE.2021.3095820
34. Astolfi, D., Marconi, L., Praly, L., Teel, A.R.: Low-power peaking-free high-gain observers. Automatica **98**, 169–179 (2018). https://doi.org/10.1016/j.automatica.2018.09.009
35. Tian, G.: Reduced-order extended state observer and frequency response analysis. Master's thesis, Cleveland State University, Cleveland (2007)
36. Skupin, P., Nowak, P., Czeczot, J.: On the stability of active disturbance rejection control for first-order plus delay time processes. ISA Trans. **125**, 179–188 (2022). https://doi.org/10.1016/j.isatra.2021.06.030
37. Łakomy, K., Madonski, R.: Cascade extended state observer for active disturbance rejection control applications under measurement noise. ISA Trans. **109**, 1–10 (2021). https://doi.org/10.1016/j.isatra.2020.09.007
38. Łakomy, K., Madonski, R., Dai, B., Yang, J., Kicki, P., Ansari, M., Li, S.: Active disturbance rejection control design with suppression of sensor noise effects in application to DC-DC buck power converter. IEEE Trans. Ind. Electron. **69**(1), 816–824 (2022). https://doi.org/10.1109/TIE.2021.3055187

References

39. Ahmad, S., Ali, A.: On active disturbance rejection control in presence of measurement noise. IEEE Trans. Ind. Electron. **69**(11), 11600–11610 (2022). https://doi.org/10.1109/TIE.2021.3121754
40. Sun, H., Madonski, R., Li, S., Zhang, Y., Xue, W.: Composite control design for systems with uncertainties and noise using combined extended state observer and Kalman filter. IEEE Trans. Ind. Electron. **69**(4), 4119–4128 (2022). https://doi.org/10.1109/TIE.2021.3075838
41. Babayomi, O., Zhang, Z., Li, Z., Heldwein, M.L., Rodriguez, J.: Robust predictive control of grid-connected converters: sensor noise suppression with parallel-cascade extended state observer. IEEE Trans. Ind. Electron. **71**(4), 3728–3740 (2023). https://doi.org/10.1109/TIE.2023.3279565
42. Linares-Flores, J., Juarez-Abad, J.A., Hernandez-Mendez, A., Castro-Heredia, O., Guerrero-Castellanos, J.F., Heredia-Barba, R., Curiel-Olivares, G.: Sliding mode control based on linear extended state observer for DC-to-DC buck–boost power converter system with mismatched disturbances. IEEE Trans. Ind. Appl. **58**(1), 940–950 (2022). https://doi.org/10.1109/TIA.2021.3130017
43. Xie, H., Li, L., Song, X., Xue, W., Song, K.: Parameter self-learning feedforward compensation-based active disturbance rejection for path-following control of self-driving forklift trucks. Asian J. Control **25**(6), 4435–4451 (2023). https://doi.org/10.1002/asjc.3110
44. Wu, Z., Shi, G., Li, D., Liu, Y., Chen, Y.: Active disturbance rejection control design for high-order integral systems. ISA Trans. **125**, 560–570 (2022). https://doi.org/10.1016/j.isatra.2021.06.038
45. Kim, Y.C., Keel, L.H., Bhattacharyya, S.P.: Transient response control via characteristic ratio assignment. IEEE Trans. Autom. Control **48**(12), 2238–2244 (2003). https://doi.org/10.1109/TAC.2003.820153
46. Grelewicz, P., Nowak, P., Czeczot, J., Musial, J.: Increment count method and its PLC-based implementation for autotuning of reduced-order ADRC with Smith predictor. IEEE Trans. Ind. Electron. **68**(12), 12554–12564 (2021). https://doi.org/10.1109/TIE.2020.3045696
47. Sun, L., Xue, W., Li, D., Zhu, H., Su, Z.: Quantitative tuning of active disturbance rejection controller for FOPTD model with application to power plant control. IEEE Trans. Ind. Electron. **69**(1), 805–815 (2022). https://doi.org/10.1109/TIE.2021.3050372
48. Kicki, P., Łakomy, K., Lee, K.M.B.: Tuning of extended state observer with neural network-based control performance assessment. Eur. J. Control. **64**, 100609 (2022). https://doi.org/10.1016/j.ejcon.2021.12.004
49. Madonski, R., Piosik, A., Herman, P.: High-gain disturbance observer tuning seen as a multicriteria optimization problem. In: Proceedings of the Mediterranean Conference on Control and Automation, pp. 1411–1416 (2013). https://doi.org/10.1109/MED.2013.6608905
50. Feng, H., Guo, B.Z.: Active disturbance rejection control: old and new results. Annu. Rev. Control. **44**, 238–248 (2017). https://doi.org/10.1016/j.arcontrol.2017.05.003
51. Zhang, X., Zhang, X., Xue, W., Xin, B.: An overview on recent progress of extended state observers for uncertain systems: methods, theory, and applications. Adv. Control Appl. **3**(2), e89 (2021). https://doi.org/10.1002/adc2.89
52. Nowicki, M., Madonski, R., Kozłowski, K.: First look at conditions on applicability of ADRC. In: Proceedings of the International Workshop on Robot Motion and Control, pp. 294–299 (2015). https://doi.org/10.1109/RoMoCo.2015.7219750
53. Chen, J., Gao, Z.: On geometric interpretation of extended state observer: a preliminary study. Control Theory Technol. **21**, 89–96 (2023). https://doi.org/10.1007/s11768-023-00130-5
54. Chen, S., Bai, W., Hu, Y., Huang, Y., Gao, Z.: On the conceptualization of total disturbance and its profound implications. Sci. China Inf. Sci. **63**(2) (2019). https://doi.org/10.1007/s11432-018-9644-3
55. Chen, W.H., Yang, J., Guo, L., Li, S.: Disturbance-observer-based control and related methods-an overview. IEEE Trans. Ind. Electron. **63**(2), 1083–1095 (2016). https://doi.org/10.1109/TIE.2015.2478397

56. Sariyildiz, E., Oboe, R., Ohnishi, K.: Disturbance observer-based robust control and its applications: 35th anniversary overview. IEEE Trans. Ind. Electron. **67**(3), 2042–2053 (2020). https://doi.org/10.1109/TIE.2019.2903752
57. Fliess, M., Join, C.: Model-free control. Int. J. Control. **86**(12), 2228–2252 (2013). https://doi.org/10.1080/00207179.2013.810345
58. Ding, S., Chen, W.H., Mei, K., Murray-Smith, D.J.: Disturbance observer design for nonlinear systems represented by input-output models. IEEE Trans. Ind. Electron. **67**(2), 1222–1232 (2020). https://doi.org/10.1109/TIE.2019.2898585
59. Czeczot, J.: Balance-based adaptive control methodology and its application to the non-isothermal CSTR. Chem. Eng. Process. Process Intensif. **45**(5), 359–371 (2006). https://doi.org/10.1016/j.cep.2005.10.002
60. Jin, H., Chen, Y., Lan, W.: Replacing PI control with first-order linear ADRC. In: Proceedings of the Data Driven Control and Learning Systems Conference, pp. 1097–1101 (2019). https://doi.org/10.1109/DDCLS.2019.8908981
61. Jin, H., Song, J., Lan, W., Gao, Z.: On the characteristics of ADRC: a PID interpretation. Sci. China Inf. Sci. **63**(10), 209201 (2020). https://doi.org/10.1007/s11432-018-9647-6
62. Li, X., Hu, Y., Gao, Z., Ai, W., Tian, S.: A PID controller based on ESO and tuning method. In: Proceedings of the Data Driven Control and Learning Systems Conference, pp. 1026–1030 (2022). https://doi.org/10.1109/DDCLS55054.2022.9858433
63. Sira-Ramirez, H., Zurita-Bustamante, E.W., Huang, C.: Equivalence among flat filters, dirty derivative-based PID controllers, ADRC, and integral reconstructor-based sliding mode control. IEEE Trans. Control Syst. Technol. **28**(5), 1696–1710 (2020). https://doi.org/10.1109/TCST.2019.2919822
64. Ahmad, S., Ali, A.: Unified disturbance-estimation-based control and equivalence with IMC and PID: case study on a DC-DC boost converter. IEEE Trans. Ind. Electron. **68**(6), 5122–5132 (2021). https://doi.org/10.1109/TIE.2020.2987269
65. Zhong, S., Huang, Y., Guo, L.: An ADRC-based PID tuning rule. Int. J. Robust Nonlinear Control, 1–14 (2021). https://doi.org/10.1002/rnc.5845
66. Huba, M., Gao, Z.: Uncovering disturbance observer and ultra-local plant models in series PI controllers. Symmetry **14**(4), 640 (2022). https://doi.org/10.3390/sym14040640
67. Stankovic, M., Ting, H., Madonski, R.: From PID to ADRC and back: expressing error-based active disturbance rejection control schemes as standard industrial 1DOF and 2DOF controllers (2023)
68. Samad, T.: A survey on industry impact and challenges thereof. IEEE Control. Syst. Mag. **37**(1), 17–18 (2017). https://doi.org/10.1109/MCS.2016.2621438
69. Han, J.: Active Disturbance Rejection Control Technique: The Technique for Estimating and Compensating the Uncertainties (in Chinese). National Defense Industry Press, Arlington (2008)
70. Li, S., Yang, J., Chen, W.H., Chen, X.: Disturbance Observer-Based Control: Methods and Applications. CRC Press, New York (2014)
71. Guo, B.Z., Zhao, Z.L.: Active Disturbance Rejection Control for Nonlinear Systems: An Introduction. Wiley, New York (2016)
72. Guo, L., Cao, S.: Anti-Disturbance Control for Systems with Multiple Disturbances. CRC Press, New York (2017)
73. Shi, D., Huang, Y., Wang, J., Shi, L.: Event-Triggered Active Disturbance Rejection Control: Theory and Applications. Springer, Berlin (2021)
74. Madonski, R., Nowicki, M., Herman, P.: Practical solution to positivity problem in water management systems—an ADRC approach. In: Proceedings of the American Control Conference, pp. 1542–1547 (2016). https://doi.org/10.1109/ACC.2016.7525135
75. Nowak, P., Stebel, K., Klopot, T., Czeczot, J., Fratczak, M., Laszczyk, P.: Flexible function block for industrial applications of active disturbance rejection controller. Arch. Control Sci. **28**, 379–400 (2018). https://doi.org/10.24425/acs.2018.124708

References

76. Miklosovic, R., Radke, A., Gao, Z.: Discrete implementation and generalization of the extended state observer. In: Proceedings of the American Control Conference, pp. 2209–2214 (2006). https://doi.org/10.1109/ACC.2006.1656547
77. Ramírez-Neria, M., Luviano-Juárez, A., Lozada-Castillo, N., Ochoa-Ortega, G., Madonski, R.: Discrete-time active disturbance rejection control: a delta operator approach. In: Advanced, Contemporary Control, pp. 1383–1395 (2020). https://doi.org/10.1007/978-3-030-50936-1_115
78. Herbst, G.: A minimum-footprint implementation of discrete-time ADRC. In: Proceedings of the European Control Conference, pp. 107–112 (2021). https://doi.org/10.23919/ECC54610.2021.9655120
79. Herbst, G., Madonski, R.: Tuning and implementation variants of discrete-time ADRC. Control Theory Technol. **21**, 72–88 (2023). https://doi.org/10.1007/s11768-023-00127-0
80. Herbst, G., Hempel, A.J., Göhrt, T., Streif, S.: Half-gain tuning for active disturbance rejection control. In: Proceedings of the IFAC World Congress, pp. 1319–1324 (2020). https://doi.org/10.1016/j.ifacol.2020.12.1864

Open Access This chapter is licensed under the terms of the Creative Commons Attribution 4.0 International License (http://creativecommons.org/licenses/by/4.0/), which permits use, sharing, adaptation, distribution and reproduction in any medium or format, as long as you give appropriate credit to the original author(s) and the source, provide a link to the Creative Commons license and indicate if changes were made.

The images or other third party material in this chapter are included in the chapter's Creative Commons license, unless indicated otherwise in a credit line to the material. If material is not included in the chapter's Creative Commons license and your intended use is not permitted by statutory regulation or exceeds the permitted use, you will need to obtain permission directly from the copyright holder.

Part II
Going Practical

Part II

Living Theories

Chapter 8
Discrete-Time Linear ADRC

Abstract We have now reached an important milestone in the book, and with this chapter we are moving from the theoretical foundations of ADRC to its applications. Putting ADRC in practice will almost always be in the form of a software-based implementation, be it in embedded systems or on PLC. Given the inherent discrete-time nature of the underlying target processor systems, it is obvious that the continuous-time variants of ADRC discussed so far are not yet the answer when asking for an actual implementation. To cross the line between "theory" and "practice" that has been established in the subtitle of this book, we will therefore present several discrete-time variants of linear ADRC with different feature sets in this chapter. All of them are ready for use in industrial practice and range from state-space to transfer function forms optimized for a low computational footprint.

8.1 State-Space Form

Based on the continuous-time state-space description of linear ADRC introduced in Chap. 3, we now want to derive a discrete-time version. This will involve controller, observer, and the tuning methods for both components.

Going from continuous to discrete time is a lossy procedure, and multiple approaches exist. They can be divided into three categories [1]: numerical integration, pole and zero mapping, and hold equivalence. For ADRC, a comparison in [2] revealed that zero-order hold (ZOH) equivalence performed best. We will follow that recommendation in this book.

Reviewing the linear ADRC equations summarized in Sect. 3.1.5, it becomes obvious that the control law (3.15) merely consists of static state feedback. Replacing the continuous-time variable t with an integer sampling instant k is therefore already sufficient to arrive at the discrete-time control law:

$$u(k) = \frac{1}{b_0} \cdot \left(k_1 \cdot r(k) - \begin{pmatrix} \boldsymbol{k}^\mathrm{T} & 1 \end{pmatrix} \cdot \hat{\boldsymbol{x}}(k) \right) \quad \text{with} \quad \boldsymbol{k}^\mathrm{T} = \begin{pmatrix} k_1 & \cdots & k_N \end{pmatrix}. \quad (8.1)$$

8.1.1 Discretization of the Observer

The only dynamic system within ADRC that requires discretization is the observer (ESO). Remember that, structurally, the ESO (3.16) is an ordinary Luenberger observer. Consequently, we can tackle the problem of discretization in a very general manner. The starting point is the continuous-time observer equation:

$$\dot{\hat{x}}(t) = A \cdot \hat{x}(t) + b \cdot u(t) + l \cdot \left(y(t) - c^T \cdot \hat{x}(t) \right). \tag{8.2}$$

The zero-order hold discretization of (8.2) delivers the following discrete-time equivalent, which is depicted in Fig. 8.1:

$$\hat{x}_p(k+1) = A_d \cdot \hat{x}_p(k) + b_d \cdot u(k) + l_p \cdot \left(y(k) - c^T \cdot \hat{x}_p(k) \right) \tag{8.3}$$

$$\text{with} \quad A_d = I + \sum_{i=1}^{\infty} \frac{A^i T_s^i}{i!}, \quad b_d = \left(\sum_{i=1}^{\infty} \frac{A^{i-1} T_s^i}{i!} \right) \cdot b, \quad c_d^T = c^T.$$

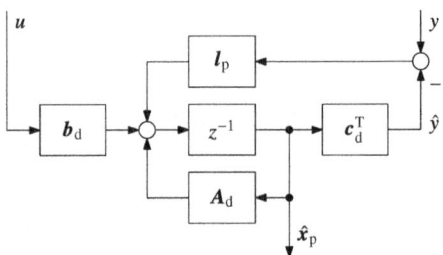

Fig. 8.1 Discrete-time predictive observer. Note the z^{-1} (delay) in the path from y to the state estimation $\hat{x}_p$, which means that the current estimate was predicted using the previous measurement

We could declare our task finished at this point, but an improvement is possible. Note from (8.2) that the most recent measurement $y(k)$ affects $\hat{x}$ at time $k + 1$: The estimate is predicted one cycle ahead. This observer is therefore also known as *predictive observer* [1], which is why we attached a "p" subscript to $\hat{x}_p$ and l_p. While "prediction" might sound good, this means that the output $u(k)$ of our control law (8.1) would be based on measurements made only up to time $k - 1$, although $y(k)$ might already be available.

As an improvement in this regard, an alternative observer variant known as *current observer* [1] can be employed—another recommendation for discrete-time ADRC put forward by Miklosovic et al. [2]. A *current observer* solves the problem of incorporating the most recent measurement $y(k)$ in the current estimate $\hat{x}(k)$ by

8.1 State-Space Form

splitting the estimation into a two-step procedure (prediction and correction), similar to a discrete-time Kalman filter. The update equations are given as follows, shown in Fig. 8.2:

$$\hat{x}(k|k-1) = A_d \cdot \hat{x}(k-1|k-1) + b_d \cdot u(k-1) \qquad \text{(prediction)},$$

$$\hat{x}(k|k) = \hat{x}(k|k-1) + l_c \cdot \left(y(k) - c_d^T \cdot \hat{x}(k|k-1) \right) \qquad \text{(correction)}.$$

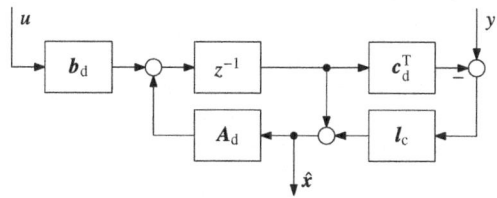

Fig. 8.2 Discrete-time current observer. Compared to the predictive observer in Fig. 8.1, the current estimate $\hat{x}$ now includes the most recent measurement y

Putting the prediction into the correction equation and abbreviating $\hat{x}(k|k) = \hat{x}(k)$ lead to a combined equation for the *current observer*, visualized in Fig. 8.3:

$$\hat{x}(k) = \left(A_d - l_c \cdot c_d^T \cdot A_d\right) \cdot \hat{x}(k-1) + \left(b_d - l_c \cdot c_d^T \cdot b_d\right) \cdot u(k-1) + l_c \cdot y(k). \quad (8.4)$$

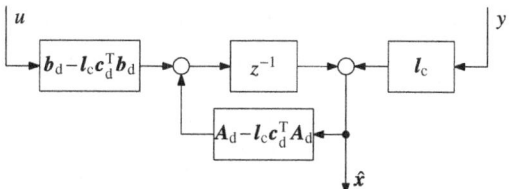

Fig. 8.3 Discrete-time current observer, condensed form obtained from Fig. 8.2 via block diagram algebra. We will use this form for our discrete-time ADRC implementation

Since the *current observer* is the preferred variant for ADRC, we will, from now on, just denote the discrete-time observer gains as $l = l_c$. Furthermore, we introduce the abbreviated matrix $A_{\text{ESO}} = A_d - l \cdot c_d^T \cdot A_d$ and vector $b_{\text{ESO}} = b_d - l \cdot c_d^T \cdot b_d$ that allow us to give a compact form of the observer equation (8.4).

For an actual (typically software-based) implementation, either Figs. 8.2 or 8.3 could be used. In this book, we will stick to the latter form, which reduces the computational footprint of the implementation in terms of the number of intermediate variables, coefficients, and arithmetic operations during runtime. As nothing comes for free, however, it should be kept in mind that a drawback of this more efficient form arises from the fact that the observer gains l are also embedded in both matrix A_{ESO} and vector b_{ESO}. Hence, when (re)tuning the observer, A_{ESO} and b_{ESO} must be updated, as well.

Concluding our derivation of the discrete-time extended state observer, we can now present its equation in compact form, which is shown as part of the block diagram of discrete-time ADRC in Fig. 8.4:

$$\hat{\boldsymbol{x}}(k) = \boldsymbol{A}_{\text{ESO}} \cdot \hat{\boldsymbol{x}}(k-1) + \boldsymbol{b}_{\text{ESO}} \cdot u(k-1) + \boldsymbol{l} \cdot y(k) \tag{8.5}$$

with $\boldsymbol{l} = \begin{pmatrix} l_1 & \cdots & l_{N+1} \end{pmatrix}^{\text{T}}$, $\boldsymbol{A}_{\text{ESO}} = \boldsymbol{A}_{\text{d}} - \boldsymbol{l} \cdot \boldsymbol{c}_{\text{d}}^{\text{T}} \cdot \boldsymbol{A}_{\text{d}}$, $\boldsymbol{b}_{\text{ESO}} = \boldsymbol{b}_{\text{d}} - \boldsymbol{l} \cdot \boldsymbol{c}_{\text{d}}^{\text{T}} \cdot \boldsymbol{b}_{\text{d}}$,

where $\boldsymbol{A}_{\text{d}} = \boldsymbol{I} + \sum_{i=1}^{\infty} \dfrac{\boldsymbol{A}^i T_s^i}{i!}$, $\boldsymbol{b}_{\text{d}} = \left(\sum_{i=1}^{\infty} \dfrac{\boldsymbol{A}^{i-1} T_s^i}{i!} \right) \cdot \boldsymbol{b}$, $\boldsymbol{c}_{\text{d}}^{\text{T}} = \boldsymbol{c}^{\text{T}}$.

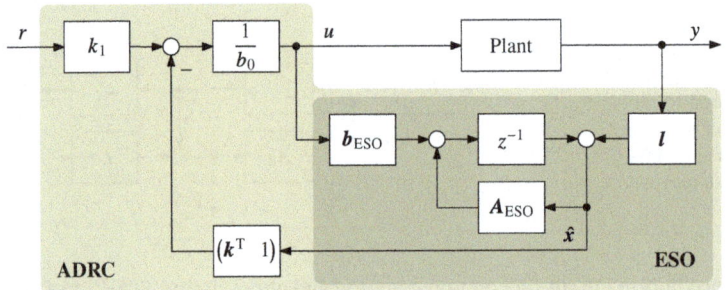

Fig. 8.4 Block diagram of a control loop with discrete-time ADRC in a state-space form, consisting of control law (8.1) and observer (8.5), which in turn is based on zero-order hold (ZOH) discretization and the *current observer* approach

For first- and second-order ADRCs, we will provide the matrices and vectors required for implementing the observer in detail:

For $N = 1$: The continuous-time $\boldsymbol{A}$ and $\boldsymbol{b}$ are

$$\boldsymbol{A} = \begin{pmatrix} 0 & 1 \\ 0 & 0 \end{pmatrix} \quad \text{and} \quad \boldsymbol{b} = \begin{pmatrix} b_0 \\ 0 \end{pmatrix}.$$

Zero-order hold discretization yields

$$\boldsymbol{A}_{\text{d}} = \begin{pmatrix} 1 & T_s \\ 0 & 1 \end{pmatrix} \quad \text{and} \quad \boldsymbol{b}_{\text{d}} = \begin{pmatrix} b_0 T_s \\ 0 \end{pmatrix}.$$

For the observer equation (8.5), we therefore obtain $\boldsymbol{A}_{\text{ESO}}$ and $\boldsymbol{b}_{\text{ESO}}$ as

$$\boldsymbol{A}_{\text{ESO}} = \begin{pmatrix} 1 - l_1 & T_s - l_1 T_s \\ -l_2 & 1 - l_2 T_s \end{pmatrix} \quad \text{and} \quad \boldsymbol{b}_{\text{ESO}} = \begin{pmatrix} b_0 T_s - l_1 b_0 T_s \\ -l_2 b_0 T_s \end{pmatrix}.$$

For $N = 2$: We similarly start with the continuous-time $\boldsymbol{A}$ and $\boldsymbol{b}$:

$$\boldsymbol{A} = \begin{pmatrix} 0 & 1 & 0 \\ 0 & 0 & 1 \\ 0 & 0 & 0 \end{pmatrix} \quad \text{and} \quad \boldsymbol{b} = \begin{pmatrix} 0 \\ b_0 \\ 0 \end{pmatrix}.$$

8.1 State-Space Form

Zero-order hold discretization yields

$$A_\mathrm{d} = \begin{pmatrix} 1 & T_\mathrm{s} & \frac{1}{2}T_\mathrm{s}^2 \\ 0 & 1 & T_\mathrm{s} \\ 0 & 0 & 1 \end{pmatrix} \quad \text{and} \quad b_\mathrm{d} = \begin{pmatrix} \frac{1}{2}b_0 T_\mathrm{s}^2 \\ b_0 T_\mathrm{s} \\ 0 \end{pmatrix}.$$

The final results A_ESO and b_ESO accordingly read

$$A_\mathrm{ESO} = \begin{pmatrix} 1 - l_1 & T_\mathrm{s} - l_1 T_\mathrm{s} & \frac{1}{2}T_\mathrm{s}^2 - \frac{1}{2}l_1 T_\mathrm{s}^2 \\ -l_2 & 1 - l_2 T_\mathrm{s} & T_\mathrm{s} - \frac{1}{2}l_2 T_\mathrm{s}^2 \\ -l_3 & -l_3 T_\mathrm{s} & 1 - \frac{1}{2}l_3 T_\mathrm{s}^2 \end{pmatrix}$$

$$\text{and} \quad b_\mathrm{ESO} = \begin{pmatrix} \frac{1}{2}b_0 T_\mathrm{s}^2 - \frac{1}{2}l_1 b_0 T_\mathrm{s}^2 \\ b_0 T_\mathrm{s} - \frac{1}{2}l_2 b_0 T_\mathrm{s}^2 \\ -\frac{1}{2}l_3 b_0 T_\mathrm{s}^2 \end{pmatrix}.$$

8.1.2 Tuning the Discrete-Time Observer

With the *current observer* approach, the error dynamics depend on the system matrix of the observer $A_\mathrm{ESO} = A_\mathrm{d} - l \cdot c_\mathrm{d}^\mathrm{T} \cdot A_\mathrm{d}$:

$$e_x(k+1) = x(k+1) - \hat{x}(k+1) = A_\mathrm{ESO} \cdot (x(k) - \hat{x}(k)).$$

Similar to the continuous-time case, we can conveniently make use of *bandwidth parameterization*, placing all eigenvalues of A_ESO at a common location z_ESO, which must be mapped from s- to z-domain before. For the Nth-order case of ADRC, $(N+1)$ observer poles must be placed:

$$\det(zI - A_\mathrm{ESO}) \stackrel{!}{=} (z - z_\mathrm{ESO})^{N+1} \quad \text{with} \quad z_\mathrm{ESO} = \mathrm{e}^{-k_\mathrm{ESO} \cdot \omega_\mathrm{CL} \cdot T_\mathrm{s}}. \quad (8.6)$$

As can be seen, (8.6) is—apart from the sampling interval T_s, of course—being fed with the same tuning parameters k_ESO and ω_CL as in the continuous-time case (3.23). It must then be solved for the observer gains $l = \begin{pmatrix} l_1 & \cdots & l_{N+1} \end{pmatrix}^\mathrm{T}$ embedded in the matrix $A_\mathrm{ESO} = A_\mathrm{d} - l \cdot c_\mathrm{d}^\mathrm{T} \cdot A_\mathrm{d}$. To make this more tangible, we will do this in detail for the especially important cases of first- and second-order ADRCs:

For $N = 1$: The tuning approach (8.6) is $\det(zI - A_\mathrm{ESO}) \stackrel{!}{=} (z - z_\mathrm{ESO})^2$. In detail, this gives

$$\det \begin{pmatrix} z - 1 + l_1 & -T_\mathrm{s} + l_1 T_\mathrm{s} \\ l_2 & z - 1 + l_2 T_\mathrm{s} \end{pmatrix} \stackrel{!}{=} z^2 - 2 z_\mathrm{ESO} z + z_\mathrm{ESO}^2,$$

$$z^2 + (l_1 + l_2 T_\mathrm{s} - 2) z + (1 - l_1) \stackrel{!}{=} z^2 - 2 z_\mathrm{ESO} z + z_\mathrm{ESO}^2. \quad (8.7)$$

By comparing the coefficients of the left- and right-hand sides of (8.7), one obtains for the observer gains l_1 and l_2

$$l_1 = 1 - z_{\text{ESO}}^2 \quad \text{and} \quad l_2 = \frac{1}{T_s} \cdot (1 - z_{\text{ESO}})^2. \tag{8.8}$$

For $N = 2$: The tuning approach (8.6) is $\det(z\mathbf{I} - \mathbf{A}_{\text{ESO}}) \stackrel{!}{=} (z - z_{\text{ESO}})^3$. We expand the right-hand side to

$$(z - z_{\text{ESO}})^3 = z^3 - 3z_{\text{ESO}} z^2 + 3z_{\text{ESO}}^2 z - z_{\text{ESO}}^3 \tag{8.9}$$

and the left-hand side to

$$\det(z\mathbf{I} - \mathbf{A}_{\text{ESO}}) = \det \begin{pmatrix} z - 1 + l_1 & -T_s + l_1 T_s & -\frac{1}{2}T_s^2 + \frac{1}{2}l_1 T_s^2 \\ l_2 & z - 1 + l_2 T_s & -T_s + \frac{1}{2}l_2 T_s^2 \\ l_3 & l_3 T_s & z - 1 + \frac{1}{2}l_3 T_s^2 \end{pmatrix} =$$

$$z^3 + \left(\frac{l_3 T_s^2}{2} + l_2 T_s + l_1 - 3\right) \cdot z^2 + \left(\frac{l_3 T_s^2}{2} - l_2 T_s - 2l_1 + 3\right) \cdot z + (l_1 - 1).$$
$$\tag{8.10}$$

By comparing the coefficients of (8.9) and (8.10), one obtains the tuning rules for the observer gains l_1, l_2, and l_3:

$$l_1 = 1 - z_{\text{ESO}}^3, \quad l_2 = \frac{3}{2T_s} \cdot (1 - z_{\text{ESO}})^2 \cdot (1 + z_{\text{ESO}}),$$
$$\text{and} \quad l_3 = \frac{1}{T_s^2} \cdot (1 - z_{\text{ESO}})^3. \tag{8.11}$$

8.1.3 Tuning the Discrete-Time Controller

The outer state feedback controller of ADRC consists of a static feedback gain vector $\mathbf{k}^\text{T}$, which means that there are no dynamics to be discretized for the controller. However, as the states of the virtual plant to be controlled by $\mathbf{k}^\text{T}$ are now being provided by a discrete-time observer, the plant dynamics should be described in discrete time, as well, to have a basis for a discrete-time design of the controller gains. We will therefore follow [3] and derive a discrete-time state-space model of the virtual plant (3.17) using zero-order hold discretization:

$$\hat{\mathbf{x}}_{\text{VP}}(k+1) = \mathbf{A}_{\text{VP,d}} \cdot \hat{\mathbf{x}}_{\text{VP}}(k) + \mathbf{b}_{\text{VP,d}} \cdot u(k), \quad y_{\text{VP}}(k) = \mathbf{c}_{\text{VP,d}}^\text{T} \cdot \hat{\mathbf{x}}_{\text{VP}}(k) \tag{8.12}$$

$$\text{with} \quad \mathbf{A}_{\text{VP,d}} = \mathbf{I} + \sum_{i=1}^{\infty} \frac{\mathbf{A}_{\text{VP}}^i T_s^i}{i!}, \quad \mathbf{b}_{\text{VP,d}} = \left(\sum_{i=1}^{\infty} \frac{\mathbf{A}_{\text{VP}}^{i-1} T_s^i}{i!}\right) \cdot \mathbf{b}_{\text{VP}}, \quad \mathbf{c}_{\text{VP,d}}^\text{T} = \mathbf{c}_{\text{VP}}^\text{T}.$$

8.1 State-Space Form

Using *bandwidth parameterization*, all poles of the desired closed-loop dynamics are being placed at the same location z_{CL} in z-domain, which gets mapped from the s-plane using $z_{CL} = e^{-\omega_{CL} T_s}$. The pole placement approach therefore reads

$$\det\left(zI - \left(A_{VP,d} - b_{VP,d} k^T\right)\right) \stackrel{!}{=} (z - z_{CL})^N \quad \text{with} \quad z_{CL} = e^{-\omega_{CL} T_s}. \quad (8.13)$$

For the practically important cases of first- and second-order ADRCs, we will give the solution to (8.13) in detail:

For $N = 1$: The elements of $A_{VP,d}$ and $b_{VP,d}$ are trivially obtained as

$$A_{VP,d} = \begin{pmatrix} 1 \end{pmatrix} \quad \text{and} \quad b_{VP,d} = \begin{pmatrix} T_s \end{pmatrix}.$$

Equation (8.13) thus becomes

$$z + (T_s k_1 - 1) \stackrel{!}{=} z - z_{CL},$$

and we can easily obtain the solution for the only controller gain k_1 as

$$k_1 = \frac{1 - z_{CL}}{T_s}. \quad (8.14)$$

Equation (8.14) is the discrete-time controller tuning rule for first-order ADRC based on *bandwidth parameterization* with $z_{CL} = e^{-\omega_{CL} T_s}$.

For $N = 2$: The elements of $A_{VP,d}$ and $b_{VP,d}$ are

$$A_{VP,d} = \begin{pmatrix} 1 & T_s \\ 0 & 1 \end{pmatrix} \quad \text{and} \quad b_{VP,d} = \begin{pmatrix} \frac{1}{2} T_s^2 \\ T_s \end{pmatrix}.$$

Equation (8.13) now reads

$$z^2 + \left(\frac{k_1 T_s^2}{2} + k_2 T_s - 2\right) \cdot z + \left(\frac{k_1 T_s^2}{2} - k_2 T_s + 1\right) \stackrel{!}{=} z^2 - 2 z_{CL} \cdot z + z_{CL}^2,$$

from which, after solving for k_1 and k_2, we obtain these controller gains as

$$k_1 = \frac{(1 - z_{CL})^2}{T_s^2} \quad \text{and} \quad k_2 = \frac{4 - (1 + z_{CL})^2}{2 T_s}. \quad (8.15)$$

8.1.4 Comparison and Summary

Discrete-Time Versus Continuous-Time Tuning

If the sampling frequencies are high enough compared to the closed-loop bandwidth, the behavior of the discrete-time virtual plant becomes indistinguishable from its continuous-time version. In this case, which can be seen as the most common one in practice, the state-feedback controller can be tuned with the continuous-time approach, leading to the already known simpler equations for $\boldsymbol{k}^\mathrm{T}$. We will demonstrate this for the first- and second-order cases in the following. Remembering the power series characterization of a real-valued exponential function, we can express z_CL as

$$z_\mathrm{CL} = e^{-\omega_\mathrm{CL} T_\mathrm{s}} = 1 + (-\omega_\mathrm{CL} T_\mathrm{s}) + \frac{(-\omega_\mathrm{CL} T_\mathrm{s})^2}{2!} + \frac{(-\omega_\mathrm{CL} T_\mathrm{s})^3}{3!} + \cdots,$$

which, when we truncate after the second term given a sufficiently small value of $\omega_\mathrm{CL} T_\mathrm{s} \ll 1$ that can be assumed for sampling frequencies well above the desired closed-loop bandwidth (suggestion for a practical rule of thumb: $\omega_\mathrm{CL} T_\mathrm{s} \lessgtr 0.05$), gets reduced to the following approximation:

$$z_\mathrm{CL} \approx 1 - \omega_\mathrm{CL} T_\mathrm{s}. \tag{8.16}$$

Putting (8.16) into (8.14) and (8.15) leads to

$$k_1 = \frac{1 - z_\mathrm{CL}}{T_\mathrm{s}} \approx \frac{1 - (1 - \omega_\mathrm{CL} T_\mathrm{s})}{T_\mathrm{s}} = \omega_\mathrm{CL} \tag{8.17}$$

for first-order ADRC, as well as

$$k_1 = \frac{(1 - z_\mathrm{CL})^2}{T_\mathrm{s}^2} \approx \frac{(1 - (1 - \omega_\mathrm{CL} T_\mathrm{s}))^2}{T_\mathrm{s}^2} = \omega_\mathrm{CL}^2, \quad \text{and}$$

$$k_2 = \frac{4 - (1 + z_\mathrm{CL})^2}{2 T_\mathrm{s}} \approx \frac{4 - (1 + (1 - \omega_\mathrm{CL} T_\mathrm{s}))^2}{2 T_\mathrm{s}} = \omega_\mathrm{CL} \cdot \left(2 - \frac{\omega_\mathrm{CL} T_\mathrm{s}}{2}\right) \approx 2 \omega_\mathrm{CL}$$

$$\tag{8.18}$$

for the second-order case, thus obtaining the tuning values of (3.19). These approximations are accurate enough for many practical cases, in which a continuous-time controller tuning can simply be transferred to its discrete-time implementation. This is shown in Fig. 8.5, where (8.17) and (8.18) are compared to (3.19). As visible, the continuous-time design leads to excessive controller gains only above $\omega_\mathrm{CL} T_\mathrm{s} \gtrsim 0.1$.

Similarly, it is possible to examine the relation between discrete- and continuous-time observer gains (when using *bandwidth parameterization*), especially since the solutions given in (8.8) and (8.11) appear to be very different to the continuous-time gains from (3.24). If $k_\mathrm{ESO} \omega_\mathrm{CL} T_\mathrm{s} \ll 1$ holds, z_ESO can approximately be expressed as

$$z_\mathrm{ESO} = e^{-k_\mathrm{ESO} \omega_\mathrm{CL} T_\mathrm{s}} = 1 + (-k_\mathrm{ESO} \omega_\mathrm{CL} T_\mathrm{s}) + \frac{(-k_\mathrm{ESO} \omega_\mathrm{CL} T_\mathrm{s})^2}{2!} + \cdots$$

$$\approx 1 - k_\mathrm{ESO} \omega_\mathrm{CL} T_\mathrm{s}. \tag{8.19}$$

For the first-order case, putting (8.19) into (8.8) and, where necessary, further simplifying by making use of the assumption $k_\mathrm{ESO} \omega_\mathrm{CL} T_\mathrm{s} \ll 1$ yield

8.1 State-Space Form

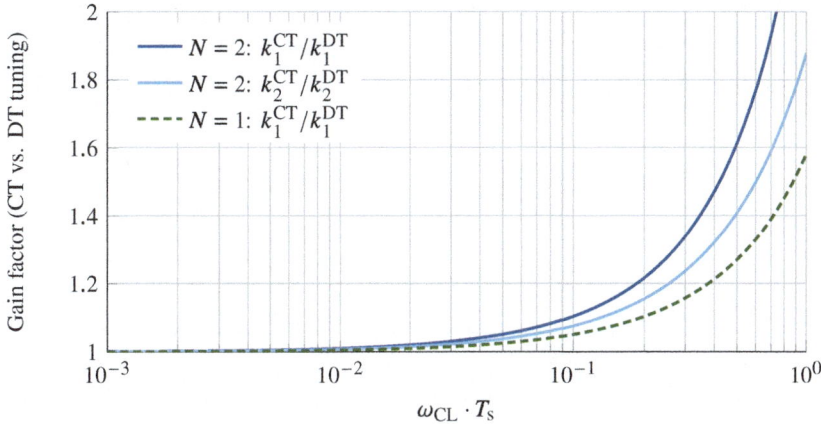

Fig. 8.5 Ratio of controller gains for first- and second-order ADRCs obtained by continuous- and discrete-time tuning. Below $\omega_{CL}T_s \lessapprox 0.05$ the sampling frequency $\frac{1}{T_s}$ becomes high enough that both designs lead to almost identical results

$$l_1 = 1 - z_{ESO}^2 \approx 1 - (1 - \omega_{CL}T_s)^2 \approx T_s \cdot 2k_{ESO}\omega_{CL}, \quad \text{and}$$

$$l_2 = \frac{1}{T_s} \cdot (1 - z_{ESO})^2 \approx \frac{1}{T_s} \cdot (1 - (1 - k_{ESO}\omega_{CL}T_s))^2 = T_s \cdot k_{ESO}^2 \omega_{CL}^2. \quad (8.20)$$

For the second-order case, we obtain

$$l_1 = 1 - z_{ESO}^3 \approx 1 - (1 - k_{ESO}\omega_{CL}T_s)^3 \approx T_s \cdot 3k_{ESO}\omega_{CL},$$

$$l_2 = \frac{3}{2T_s} \cdot (1 - z_{ESO})^2 \cdot (1 + z_{ESO})$$

$$\approx \frac{3}{2T_s} \cdot (k_{ESO}\omega_{CL}T_s)^2 \cdot (2 - k_{ESO}\omega_{CL}T_s) \approx T_s \cdot 3k_{ESO}^2\omega_{CL}^2,$$

$$l_3 = \frac{1}{T_s^2} \cdot (1 - z_{ESO})^3 \approx \frac{1}{T_s^2} \cdot (1 - (1 - k_{ESO}\omega_{CL}T_s))^3 = T_s \cdot k_{ESO}^3 \omega_{CL}^3. \quad (8.21)$$

In both cases ($N = 1$ and $N = 2$), the discrete-time observer gains are approximately related to the continuous-time gains from (3.24) by the factor T_s, if the latter is small enough to warrant the assumption $k_{ESO}\omega_{CL}T_s \ll 1$. Figure 8.6 compares discrete- and continuous-time observer gains also for cases where this assumption is being violated. The scaling factor T_s can be attributed to the replacement of the continuous-time integrator by an accumulator in the discrete-time observer.

Let us summarize this comparison of continuous- and discrete-time controllers and observer tuning. We can state that, while it is possible to reuse the continuous-time gains for a discrete-time observer implementation, we do not recommend doing so. For the observer design, it is much easier to violate the assumption $k_{ESO}\omega_{CL}T_s \ll 1$ due to the relative bandwidth factor k_{ESO}. We therefore only use the exact discrete-time observer gains in this book. And although $\omega_{CL}T_s \ll 1$ will probably be fulfilled in most practical cases, we default to using exact solutions of the discrete-time pole placement for the outer control loop in this book, as well. When deriving transfer

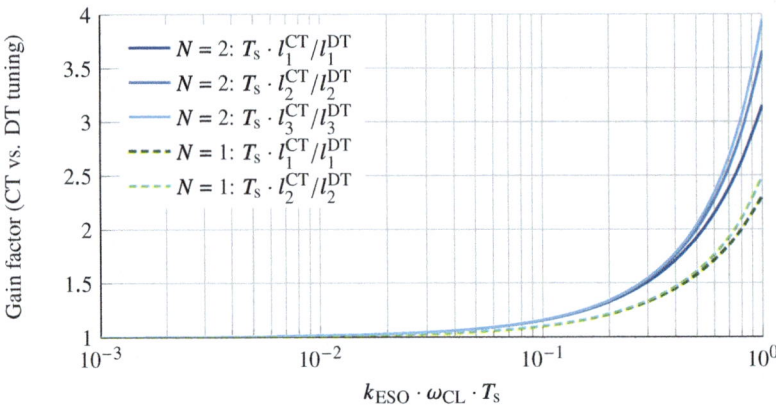

Fig. 8.6 Ratio of observer gains for first- and second-order ADRCs obtained by continuous- and discrete-time tuning. To be comparable, the continuous-time gains must be multiplied by the width of the sampling interval T_s, as the integrator of the continuous-time observer is replaced by an accumulator in the discrete-time case. Similar to Fig. 8.5, one can argue that below $k_{ESO}\omega_{CL}T_s \lesssim 0.05$ the sampling frequency $\frac{1}{T_s}$ becomes high enough that the continuous-time tuning can be reused for a discrete-time implementation. Yet one must note that, in contrast to comparing ω_{CL} to $\frac{1}{T_s}$ in Fig. 8.5, the additional relative bandwidth factor k_{ESO} has to be taken into account. It is therefore much more likely to violate $k_{ESO}\omega_{CL}T_s \lesssim 0.05$ during observer tuning than $\omega_{CL}T_s \lesssim 0.05$ for the outer control loop. As the mismatch of continuous- and discrete-time gains is even more pronounced in this case, we recommend always using the exact discrete-time observer gains

function forms of discrete-time ADRC in subsequent sections, we will, however, always provide the option to use a custom tuning such as the continuous-time gains as an alternative.

Bottom Line

Incorporating the elements discussed so far in this chapter, related to discrete observer and controller and their tuning, we are ready to present our first "cooking recipe" for discrete-time ADRC, here in its state-space form.

> **Cooking Recipe (Discrete-Time ADRC, State-Space Form)**
>
> To implement and tune the discrete-time state-space variant of ADRC that we have derived in this section, the following steps have to be carried out:
>
> 1. *Plant modeling:* Identify the order N and the gain parameter b_0 of the plant model.
> 2a. *Controller tuning:* When using *bandwidth parameterization*, the tuning parameter is ω_{CL}. The N controller gains $k_{1,...,N}$ for a discrete-

(continued)

8.2 Transfer Function Form

time implementation with sampling interval T_s are obtained by solving (8.13). For sufficiently small sampling intervals $\omega_{CL} T_s \ll 1$, which is the most likely case, it is also possible to simply use the continuous-time controller tuning (3.19).

2b. *Observer tuning:* For *bandwidth parameterization*, the additional tuning parameter k_{ESO} (relative observer bandwidth) must be chosen. The $(N+1)$ observer gains $l_{1,\ldots,N+1}$ are then obtained by solving (8.6).

3. *Implementation:* The discrete-time state-space form of ADRC consists of the control law (8.1) and the observer equation (8.5)—to be implemented as shown in Fig. 8.4.

4. *Customization:* A practical controller will at least require an add-on mechanism for control signal limitation and windup protection, which we will cover in Sect. 9.1. We can already tell that this variant is well prepared. If a limited control signal value u_{lim} is computed outside of the control law, windup can be avoided by simply feeding the value $u_{lim}(k-1)$ instead of $u(k-1)$ back into the observer (8.5).

All details required for implementing and tuning ADRC are once more compiled in the Appendix, for the discrete-time state-space variant in Appendix A.5.

8.2 Transfer Function Form

ADRC lines up as a general-purpose controller, an alternative to classical PI or PID controllers. These will almost always *not* be implemented in a state-space form, as done for ADRC in the previous section. This could present a barrier to adopting ADRC for several reasons:

- People are more familiar with transfer function forms of controllers and find it difficult to draw comparisons.
- (Software) Building blocks exist for a transfer function controller implementation and cannot be reused.
- At least for $N \geq 2$, the matrix multiplication in the observer (8.5) comes with a higher computational footprint than a typical discrete-time transfer function implementation of a controller.

In this section we will therefore—as done in [4]—introduce a transfer function form of discrete-time ADRC that exhibits great similarity to "classical" controller implementations, addressing all of the problems mentioned above.

Derivation

In contrast to a state-space form, transfer functions directly relate the input signals r and y of the controller to its output signal u. When deriving a transfer function representation, the state variables $\hat{x}$ will therefore not be available anymore.

To obtain discrete-time transfer functions, we start by transforming the time-domain equations of controller (8.1) and observer (8.5) into z-domain:

$$u(z) = \frac{1}{b_0} \cdot \left(k_1 \cdot r(z) - \left(\boldsymbol{k}^\mathrm{T}\ 1\right) \cdot \hat{\boldsymbol{x}}(z)\right), \tag{8.22}$$

$$\hat{\boldsymbol{x}}(z) = z^{-1} \cdot \boldsymbol{A}_{\mathrm{ESO}} \cdot \hat{\boldsymbol{x}}(z) + z^{-1} \cdot \boldsymbol{b}_{\mathrm{ESO}} \cdot u(z) + \boldsymbol{l} \cdot y(z). \tag{8.23}$$

If we put (8.22) into (8.23), we obtain the closed-loop dynamics of the discrete-time observer, introducing the matrix $\boldsymbol{\Phi}_{\mathrm{ESO}}$ as a shorthand notation:

$$\hat{\boldsymbol{x}}(z) = \boldsymbol{\Phi}_{\mathrm{ESO}} \cdot \left(z^{-1} \cdot \frac{k_1}{b_0} \cdot \boldsymbol{b}_{\mathrm{ESO}} \cdot r(z) + \boldsymbol{l} \cdot y(z)\right) \tag{8.24}$$

$$\text{with}\quad \boldsymbol{\Phi}_{\mathrm{ESO}} = \left(\boldsymbol{I} - z^{-1} \cdot \left(\boldsymbol{A}_{\mathrm{ESO}} - \frac{1}{b_0} \cdot \boldsymbol{b}_{\mathrm{ESO}} \cdot \left(\boldsymbol{k}^\mathrm{T}\ 1\right)\right)\right)^{-1}. \tag{8.25}$$

This enables us to eliminate $\hat{\boldsymbol{x}}(z)$ by putting (8.24) back into (8.22). The control law including the closed-loop observer dynamics now reads

$$u(z) = \frac{k_1}{b_0} \cdot r(z) - \frac{1}{b_0} \cdot \left(\boldsymbol{k}^\mathrm{T}\ 1\right) \cdot \boldsymbol{\Phi}_{\mathrm{ESO}} \cdot \left(z^{-1} \cdot \frac{k_1}{b_0} \cdot \boldsymbol{b}_{\mathrm{ESO}} \cdot r(z) + \boldsymbol{l} \cdot y(z)\right) \tag{8.26}$$

$$= \frac{k_1}{b_0} \cdot \left(1 - z^{-1} \cdot \left(\boldsymbol{k}^\mathrm{T}\ 1\right) \cdot \boldsymbol{\Phi}_{\mathrm{ESO}} \cdot \frac{1}{b_0} \cdot \boldsymbol{b}_{\mathrm{ESO}}\right) \cdot r(z)$$

$$- \left(\frac{1}{b_0} \cdot \left(\boldsymbol{k}^\mathrm{T}\ 1\right) \cdot \boldsymbol{\Phi}_{\mathrm{ESO}} \cdot \boldsymbol{l}\right) \cdot y(z).$$

We could already read off two transfer functions relating r and y to u from (8.26) and declare this finished, but we do not. For a good reason, which will be uncovered soon, things need to be made more complicated at first by factoring out the transfer function from $y(z)$ to $u(z)$ in (8.26). This yields

$$u(z) = \left(\frac{1}{b_0} \cdot \left(\boldsymbol{k}^\mathrm{T}\ 1\right) \cdot \boldsymbol{\Phi}_{\mathrm{ESO}} \cdot \boldsymbol{l}\right) \cdot \left[\frac{\frac{k_1}{b_0} \cdot \left(1 - z^{-1} \cdot \left(\boldsymbol{k}^\mathrm{T}\ 1\right) \cdot \boldsymbol{\Phi}_{\mathrm{ESO}} \cdot \frac{1}{b_0} \cdot \boldsymbol{b}_{\mathrm{ESO}}\right)}{\frac{1}{b_0} \cdot \left(\boldsymbol{k}^\mathrm{T}\ 1\right) \cdot \boldsymbol{\Phi}_{\mathrm{ESO}} \cdot \boldsymbol{l}} \cdot r(z) - y(z)\right]. \tag{8.27}$$

Equation (8.27) very much resembles a typical discrete-time controller structure with two transfer functions, consisting of a feedback controller $C_{\mathrm{PF}}(z)$ and a prefilter $C_{\mathrm{PF}}(z)$, shown in Fig. 8.7:

8.2 Transfer Function Form

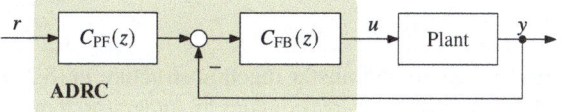

Fig. 8.7 Discrete-time transfer function form of ADRC

$$u(z) = C_{FB}(z) \cdot [C_{PF}(z) \cdot r(z) - y(z)]. \tag{8.28}$$

Comparing (8.27) and (8.28), we can now read off the transfer functions of the feedback controller $C_{FB}(z)$ and the prefilter $C_{PF}(z)$:

$$C_{FB}(z) = \frac{1}{b_0} \cdot (\mathbf{k}^T \ 1) \cdot \boldsymbol{\Phi}_{ESO} \cdot \mathbf{l}, \tag{8.29}$$

$$C_{PF}(z) = \frac{\frac{k_1}{b_0} \cdot \left(1 - z^{-1} \cdot (\mathbf{k}^T \ 1) \cdot \boldsymbol{\Phi}_{ESO} \cdot \frac{1}{b_0} \cdot \mathbf{b}_{ESO}\right)}{\frac{1}{b_0} \cdot (\mathbf{k}^T \ 1) \cdot \boldsymbol{\Phi}_{ESO} \cdot \mathbf{l}}. \tag{8.30}$$

Remember from the continuous-time transfer function representation developed in Sect. 4.2 that the feedback controller includes an integrator. Consequently $C_{FB}(z)$ will feature a discrete-time integrator pole at $z = 1$. It is not strictly required, but beneficial to factor this pole out of $C_{FB}(z)$:

$$C_{FB}(z) = \Delta C_{FB}(z) \cdot \frac{1}{1 - z^{-1}}. \tag{8.31}$$

That way, $C_{FB}(z)$ can be implemented as a series connection of a discrete-time filter and an accumulator, which has several advantages:

- The structure of (8.31) better matches its continuous-time counterpart $C_{FB}(s)$ in (4.22), where the integrator was factored out, as well.
- A superfluous (dependent) coefficient of the denominator polynomial of $\Delta C_{FB}(z)$ gets eliminated, saving one parameter to store and one multiplication at runtime.
- The accumulator in (8.31) can be easily implemented as a clamped integrator, providing simple but quite effective windup protection—$\Delta C_{FB}(z)$ has been brought into to the so-called *incremental* or *velocity* form [5, 6].

The two transfer functions can now be represented as given in (8.32). Their coefficients result from putting controller and observer gains in (8.29) and (8.30). Ready-to-use equations for α, β, γ are listed in Appendix A.6. We factored out $\frac{1}{\beta_0}$ in $C_{PF}(z)$, enabling reuse of the β and more compact equations for the γ coefficients.

$$C_{FB}(z) = \frac{\sum_{i=0}^{N} \beta_i z^{-i}}{1 + \sum_{i=1}^{N} \alpha_i z^{-i}} \cdot \frac{1}{1 - z^{-1}} \quad \text{and} \quad C_{PF}(z) = \frac{\frac{1}{\beta_0} \sum_{i=0}^{N+1} \gamma_i z^{-i}}{1 + \frac{1}{\beta_0} \sum_{i=1}^{N} \beta_i z^{-i}}. \tag{8.32}$$

Discussion

We can now comment on our preference of the transfer function structure in (8.27) to the one of (8.26). As the feedback controller $C_{\text{FB}}(z)$ contains an integrator, it must only be fed with the control error. This ensures that the integrator remains constant in steady state when the control error is eliminated. In the structure of (8.26), both transfer functions would contain an integrator, their integrator states running away to infinity for any nonzero reference r. Not a problem in theory, but very much so in practice!

In contrast to the continuous-time case in Sect. 4.2, no realizability issues for $C_{\text{PF}}(z)$ arise from the fact that its numerator polynomial is of higher order than its denominator. The introduction of a third (feedforward) transfer function is not necessary in the discrete-time domain.

If the runtime performance of a controller implementation is important, this transfer function form is more efficient than the state-space variant of Sect. 8.1. The number of multiplications and additions only scales linearly with N here, compared to a quadratic growth of operations in the state-space case [7].

Compared to the state-space form, the possible feedback path of a limited control signal value was removed while deriving the transfer functions, resulting in less flexibility regarding control signal limitation and anti-windup performance. In Sect. 8.3, we will introduce a third discrete-time form, combining the efficiency of transfer functions with the flexibility of the state-space form.

Cooking Recipe (Discrete-Time ADRC, Transfer Function Form)

To implement and tune the discrete-time transfer function form of ADRC derived in this section, the following steps have to be carried out:

1. *Plant modeling:* Identify the order N and the gain parameter b_0 of the plant model.
2. *Controller and observer tuning:* When using *bandwidth parameterization*, the tuning parameters are ω_{CL} and k_{ESO}. For first- and second-order ADRCs, these values (along with the sampling interval T_s and the plant gain b_0) can simply be put in the design equations for the α, β, γ coefficients tabulated in Appendix A.6. It is also possible to compute these coefficients from the otherwise obtained state-space controller and observer gains k^{T} and l.
3. *Implementation:* The structure of the control law (8.28), shown in Fig. 8.7, is well known from existing control engineering practice. It comprises a reference signal prefilter $C_{\text{PF}}(z)$ and a feedback controller $C_{\text{FB}}(z)$, preferably with a factored-out integrator as in (8.32), which should be implemented with output limits.

As always in this book, all relevant equations and details are once more compiled in the Appendix, for this discrete-time transfer function form in Appendix A.6.

8.3 Dual-Feedback Transfer Function Form

When the observer dynamics were folded into transfer functions during the derivation in Sect. 8.2, the feedback path of the controller output back into the observer was removed. In this section, we will present an alternative transfer function form of discrete-time ADRC introduced in [7]. It preserves the feedback path of the previous controller output $u(k-1)$. This allows to apply an arbitrary control signal limitation outside of the actual controller and feed the limited signal $u_{\text{lim}}(k-1)$ back, as in the state-space variant. At the same time, it will enable an even more efficient implementation than the transfer function variant of Sect. 8.2.

Derivation

Let us once more start by transforming the discrete-time controller (8.1) and observer (8.5) into z-domain:

$$u(z) = \frac{1}{b_0} \cdot \left(k_1 \cdot r(z) - \begin{pmatrix}\boldsymbol{k}^T & 1\end{pmatrix} \cdot \hat{\boldsymbol{x}}(z)\right), \tag{8.33}$$

$$\hat{\boldsymbol{x}}(z) = z^{-1} \cdot \boldsymbol{A}_{\text{ESO}} \cdot \hat{\boldsymbol{x}}(z) + z^{-1} \cdot \boldsymbol{b}_{\text{ESO}} \cdot u(z) + \boldsymbol{l} \cdot y(z). \tag{8.34}$$

We solve (8.34) for $\hat{\boldsymbol{x}}(z)$:

$$\hat{\boldsymbol{x}}(z) = \left(\boldsymbol{I} - z^{-1}\boldsymbol{A}_{\text{ESO}}\right)^{-1} \cdot \left(z^{-1} \cdot \boldsymbol{b}_{\text{ESO}} \cdot u(z) + \boldsymbol{l} \cdot y(z)\right). \tag{8.35}$$

Equation (8.35) relates the observer inputs $u(z)$ and $y(z)$ to its output, the estimation $\hat{\boldsymbol{x}}(z)$. We have already mentioned multiple times that it will be beneficial to feed a limited control signal value $u_{\text{lim}}(z)$ into the observer to prevent possible windup issues in case a control signal limitation is being used (which one should). Replacing $u(z)$ with $u_{\text{lim}}(z)$ in (8.35) and then putting this equation into (8.33) result in

$$u(z) = \frac{1}{b_0} \cdot \left(k_1 \cdot r(z) - \begin{pmatrix}\boldsymbol{k}^T & 1\end{pmatrix} \cdot \left(\boldsymbol{I} - z^{-1}\boldsymbol{A}_{\text{ESO}}\right)^{-1} \cdot \left(z^{-1} \cdot \boldsymbol{b}_{\text{ESO}} \cdot u_{\text{lim}}(z) + \boldsymbol{l} \cdot y(z)\right)\right). \tag{8.36}$$

At this point we have obtained a control law consisting of three transfer functions (one of them being only a gain factor of $\frac{k_1}{b_0}$) that relate the controller inputs $r(z)$, $y(z)$, and $u_{\text{lim}}(z)$ to the controller output $u(z)$, which we can also write as follows:

$$u(z) = \frac{k_1}{b_0} \cdot r(z) - C_{\text{FBy}}(z) \cdot y(z) + C_{\text{FBu}}(z) \cdot u_{\text{lim}}(z) \tag{8.37}$$

$$\text{with} \quad C_{\text{FBy}}(z) = \frac{1}{b_0} \cdot \begin{pmatrix}\boldsymbol{k}^T & 1\end{pmatrix} \cdot \left(\boldsymbol{I} - z^{-1}\boldsymbol{A}_{\text{ESO}}\right)^{-1} \cdot \boldsymbol{l} \tag{8.38}$$

$$\text{and} \quad C_{\text{FBu}}(z) = -\frac{z^{-1}}{b_0} \cdot \begin{pmatrix}\boldsymbol{k}^T & 1\end{pmatrix} \cdot \left(\boldsymbol{I} - z^{-1}\boldsymbol{A}_{\text{ESO}}\right)^{-1} \cdot \boldsymbol{b}_{\text{ESO}}. \tag{8.39}$$

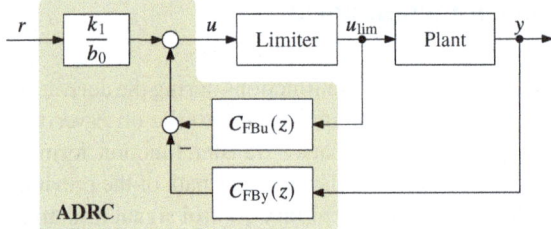

Fig. 8.8 Control loop with a dual-feedback transfer function representation of discrete-time ADRC

The two feedback transfer functions $C_{\text{FBy}}(z)$ and $C_{\text{FBu}}(z)$ of the control law (8.37), which is shown as a block diagram in Fig. 8.8, exhibit the following structure:

$$C_{\text{FBy}}(z) = \frac{\sum_{i=0}^{N} \beta_i z^{-i}}{1 + \sum_{i=1}^{N+1} \alpha_i z^{-i}} \quad \text{and} \quad C_{\text{FBu}}(z) = z^{-1} \cdot \frac{\sum_{i=0}^{N} \gamma_i z^{-i}}{1 + \sum_{i=1}^{N+1} \alpha_i z^{-i}}. \tag{8.40}$$

Their coefficients are obtained by putting controller and observer gains $\boldsymbol{k}^T$ and $\boldsymbol{l}$ in (8.38) and (8.39). Ready-to-use equations for α, β, γ are listed in Appendix A.7.

Discussion

In a "classical" control loop, there is only one transfer function, feeding back the control error. The structure with dual feedback of $u_{\text{lim}}(z)$ and $y(z)$ in this ADRC form led us to reflect this very fact in its name. Dedicated feedback of $u_{\text{lim}}(z)$ is what ensures the exact same flexibility and behavior compared to the state-space form regarding control signal limitation and windup protection.

If no control signal limitation would be used, $u_{\text{lim}}(z) = u(z)$ would be fed back to the updated controller output $u(z)$. However, in any case, no algebraic loop is being created here, as there is a one-step delay (z^{-1}) in $C_{\text{FBu}}(z)$.

Note that in (8.37) $C_{\text{FBy}}(z) \cdot y(z)$ is being fed back with a negative sign, whereas we chose positive feedback for $C_{\text{FBu}}(z) \cdot u(z)$. The latter is a sign change compared to the original derivation in [7], but for a good reason which we will uncover below. Both transfer functions $C_{\text{FBy}}(z)$ and $C_{\text{FBu}}(z)$ share the same denominator. Assuming *bandwidth parameterization*, we can express (8.38) and (8.39) as

$$C_{\text{FBy}}(z) = \frac{1}{b_0} \cdot \frac{(\boldsymbol{k}^T \; 1) \cdot \text{adj}\left(\boldsymbol{I} - z^{-1}\boldsymbol{A}_{\text{ESO}}\right) \cdot \boldsymbol{l}}{\left(1 - z^{-1} z_{\text{ESO}}\right)^{N+1}},$$

$$C_{\text{FBu}}(z) = -\frac{z^{-1}}{b_0} \cdot \frac{(\boldsymbol{k}^T \; 1) \cdot \text{adj}\left(\boldsymbol{I} - z^{-1}\boldsymbol{A}_{\text{ESO}}\right) \cdot \boldsymbol{b}_{\text{ESO}}}{\left(1 - z^{-1} z_{\text{ESO}}\right)^{N+1}}.$$

8.3 Dual-Feedback Transfer Function Form

One can now clearly see that their denominator is the characteristic polynomial of the discrete-time observer dynamics as designed in the tuning approach (8.6). This means that both transfer functions have all their poles at $z_\text{ESO} = e^{-k_\text{ESO} \cdot \omega_\text{CL} \cdot T_s}$.

So—where is the integrator?

As it turns out, $C_\text{FBu}(z)$ has a steady-state gain of $C_\text{FBu}(z \to 1) = 1$. Look at Fig. 8.8 once more. Closing the inner loop around the control signal limiter with positive feedback and $C_\text{FBu}(z \to 1) = 1$ means the inner loop acts as a limited accumulator with filtered feedback. Without a limiter, its transfer function is

$$\frac{1}{1 - C_\text{FBu}(z)} = \frac{1}{1 - z^{-1} \cdot \underbrace{\left(-\frac{1}{b_0} \cdot (\boldsymbol{k}^\mathrm{T}\ 1) \cdot \left(\boldsymbol{I} - z^{-1} \boldsymbol{A}_\text{ESO} \right)^{-1} \cdot \boldsymbol{b}_\text{ESO} \right)}_{\text{filter with unity steady-state gain}}},$$

from which the similarity with a discrete-time accumulator $\frac{1}{1-z^{-1}}$ should become obvious. Clearly indicating this behavior was our reason to use positive feedback for $C_\text{FBu}(z) \cdot u(z)$ in the control law (8.37).

Efficient Implementation

With two transfer functions, the dual-feedback variant of discrete-time ADRC is already similarly efficient as the transfer function approach from Sect. 8.2, while maintaining the exact behavior and features of the state-space form from Sect. 8.1. Yet, due to the identical denominators of $C_\text{FBy}(z)$ and $C_\text{FBu}(z)$, the implementation can be made even more efficient by making use of superposition:

$$u(z) = \frac{k_1}{b_0} \cdot r(z) + \frac{-\left(\sum_{i=0}^{N} \beta_i z^{-i} \right) \cdot y(z) + z^{-1} \cdot \left(\sum_{i=0}^{N} \gamma_i z^{-i} \right) \cdot u_\text{lim}(z)}{1 + \sum_{i=1}^{N+1} \alpha_i z^{-i}}. \quad (8.41)$$

And finally, as proposed in [7], the number of required storage variables at runtime can be made as low as for the state-space form by choosing a (transposed) "direct form II," known from digital signal processing [8]. That way, an implementation with both a minimum number of storage variables and algebraic operations of all known ADRC variants to date is obtained, which was accordingly referred to as "minimum-footprint" form in [7]. For first- and second-order ADRCs, the resulting structure of the controller is shown in Fig. 8.9. The above derivation and deliberation can be summarized with the following "cooking recipe."

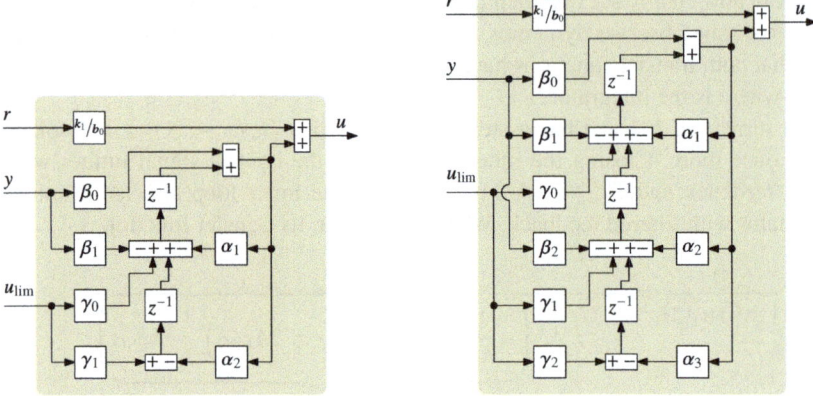

Fig. 8.9 Minimum-footprint form of first- and second-order discrete-time ADRCs based on the dual-feedback transfer function approach

---

Cooking Recipe (Discrete-Time ADRC, Dual-Feedback TF Form)

To implement and tune this alternative discrete-time transfer function form, the following steps have to be carried out:

1. *Plant modeling:* Identify the order N and the gain parameter b_0 of the plant model.
2. *Controller and observer tuning:* For *bandwidth parameterization*, the tuning parameters are ω_{CL} and k_{ESO}. Together with the sampling interval T_s and the plant gain b_0, these values can simply be put in design equations for the α, β, γ coefficients tabulated in Appendix A.7 for the first- and second-order cases. It is also possible to compute these coefficients from the otherwise obtained state-space controller and observer gains k^T and l.
3. *Implementation:* The controller structure of this dual-feedback transfer function variant is shown in Fig. 8.8. The control law (8.37) can either be implemented using the two separate transfer functions $C_{\text{FBy}}(z)$ and $C_{\text{FBu}}(z)$ from (8.40) or in the more efficient form (8.41) shown in Fig. 8.9.
4. *Customization:* As in the state-space form, an arbitrary control signal limitation add-on block can be added to a control loop with this ADRC variant, computing u_{lim} from u. Feeding back u_{lim} through the $C_{\text{FBu}}(z)$ transfer function is a simple yet effective measure of preventing windup issues.

8.4 Discrete-Time Error-Based ADRC

As for all other ADRC variants in this book, all relevant equations and details are compiled in the Appendix, for this dual-feedback transfer function form in Appendix A.7.

8.4 Discrete-Time Error-Based ADRC

Three discrete-time forms of ADRC have been derived in Sects. 8.1, 8.2, and 8.3 of this chapter. We will explicitly denote these as "output-based" ADRC variants in this section, as we are about to develop the same choices for the error-based form of linear ADRC that was introduced in Sect. 6.4.

The strong connections between output- and error-based ADRCs, previously found in Sect. 6.4, also apply to their discrete-time forms. This allows deriving the discrete-time error-based ADRC variants in a very concise manner, as done in [3]. It will be seen that the exact matrices, vectors, tuning gains, transfer functions, and coefficients of output-based ADRC can be reused. Therefore, we will refrain from giving "cooking recipes" for each variant but conclude with one common recipe, as only the controller structure will change, but not the required parameters.

8.4.1 State-Space Form

Recall the state-space derivation of discrete-time output-based ADRC in Sect. 8.1. Since the control law of ADRC is a static feedback of estimated states, providing a discrete-time equation is trivial. What has to be discretized, however, is the dynamics of the extended state observer providing these estimated states. If we compare the continuous-time state-space block diagrams of output- and error-based ADRCs in Figs. 3.5 and 6.12, it becomes obvious that the observers are very similar. The only two differences for error-based ADRC are that an input $e(t)$ instead of $y(t)$ is being used and that the controller output $u(t)$—or its limited value $u_{\lim}(t)$—enters the observer with a negative sign.

Without repeating the discretization procedure from Sect. 8.1, we can therefore directly present the discrete-time counterparts to the continuous-time equations of error-based ADRC (controller (6.27) and observer (6.26)):

$$u(k) = \frac{1}{b_0} \cdot \begin{pmatrix} \boldsymbol{k}^\mathrm{T} & 1 \end{pmatrix} \cdot \hat{\boldsymbol{x}}(k) \quad \text{with} \quad \boldsymbol{k}^\mathrm{T} = \begin{pmatrix} k_1 & \cdots & k_N \end{pmatrix}, \tag{8.42}$$

$$\hat{\boldsymbol{x}}(k) = \boldsymbol{A}_{\mathrm{ESO}} \cdot \hat{\boldsymbol{x}}(k-1) - \boldsymbol{b}_{\mathrm{ESO}} \cdot u_{\lim}(k-1) + \boldsymbol{l} \cdot e(k) \quad \text{with} \quad \boldsymbol{l} = \begin{pmatrix} l_1 & \cdots & l_{N+1} \end{pmatrix}^\mathrm{T}. \tag{8.43}$$

Comparing the resulting discrete-time observer equations for output- and error-based ADRC, (8.5) and (8.43), the only two differences are once again: input $e(k)$ instead of $y(k)$ and negative sign for $u_{\lim}(k-1)$ in (8.43). This is also apparent by comparing the block diagram in Fig. 8.10 with the output-based version in Fig. 8.4.

All matrices and vectors required to implement controller and observer, as well as their tuning methods, are identical to the output-based case. They can be found in Tables A.5.1 and A.5.2 of Appendix A.5.

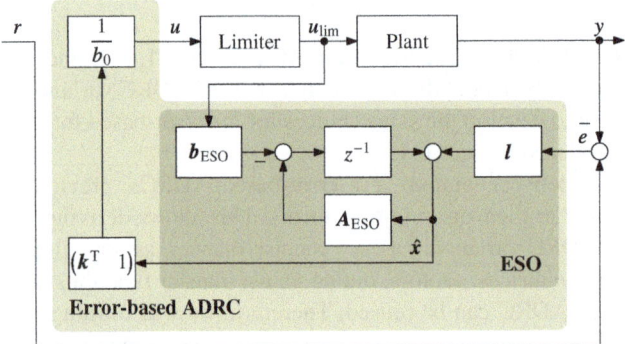

Fig. 8.10 Discrete-time error-based ADRC, state-space form, with an external control signal limiter

8.4.2 Transfer Function Form

Now that discrete-time error-based ADRC was introduced in state space, a transfer function form can be derived in the same manner as done in Sect. 8.2 for output-based ADRC.

The z-domain representation of the error-based control law (8.42) and observer (8.43) form the starting point. As seen in Sect. 8.2, the feedback path of the limited controller output will be removed during the transfer function derivation. In (8.43), we therefore have to set $u(k-1) = u_{\lim}(k-1)$ before applying the z-transform:

$$u(z) = \frac{1}{b_0} \cdot \begin{pmatrix} \boldsymbol{k}^\mathrm{T} & 1 \end{pmatrix} \cdot \hat{\boldsymbol{x}}(z), \tag{8.44}$$

$$\hat{\boldsymbol{x}}(z) = z^{-1} \cdot \boldsymbol{A}_{\mathrm{ESO}} \cdot \hat{\boldsymbol{x}}(z) - z^{-1} \cdot \boldsymbol{b}_{\mathrm{ESO}} \cdot u(z) + \boldsymbol{l} \cdot e(z). \tag{8.45}$$

Inserting (8.44) into (8.45) yields the closed-loop observer equation:

$$\hat{\boldsymbol{x}}(z) = \left(\boldsymbol{I} - z^{-1} \cdot \left(\boldsymbol{A}_{\mathrm{ESO}} - \frac{1}{b_0} \cdot \boldsymbol{b}_{\mathrm{ESO}} \cdot \begin{pmatrix} \boldsymbol{k}^\mathrm{T} & 1 \end{pmatrix} \right) \right)^{-1} \cdot \boldsymbol{l} \cdot e(z),$$

which is finally put back into (8.44):

$$u(z) = \underbrace{\frac{1}{b_0} \cdot \begin{pmatrix} \boldsymbol{k}^\mathrm{T} & 1 \end{pmatrix} \cdot \left(\boldsymbol{I} - z^{-1} \cdot \left(\boldsymbol{A}_{\mathrm{ESO}} - \frac{1}{b_0} \cdot \boldsymbol{b}_{\mathrm{ESO}} \cdot \begin{pmatrix} \boldsymbol{k}^\mathrm{T} & 1 \end{pmatrix} \right) \right)^{-1} \cdot \boldsymbol{l}}_{C_{\mathrm{FB}}(z)} \cdot e(z). \tag{8.46}$$

8.4 Discrete-Time Error-Based ADRC

Equation (8.46) is a control law that makes use of exactly the same transfer function $C_{FB}(z)$ as derived in (8.29) for output-based ADRC, and we can therefore express it as

$$u(z) = C_{FB}(z) \cdot e(z) \quad \text{with} \quad C_{FB}(z) = \frac{\sum_{i=0}^{N} \beta_i z^{-i}}{1 + \sum_{i=1}^{N} \alpha_i z^{-i}} \cdot \frac{1}{1 - z^{-1}}. \tag{8.47}$$

The α and β coefficients in (8.46) can consequently be found in the reference material of the output-based variant in Appendix A.6, specifically in Tables A.6.2 and A.6.3.

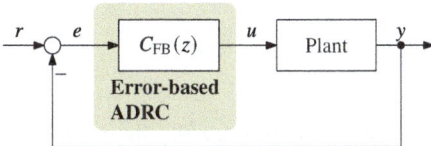

Fig. 8.11 Discrete-time error-based ADRC, transfer function form

As shown in Fig. 8.11, this form of error-based ADRC requires only one transfer function, equivalent to its continuous-time counterpart introduced in Sect. 6.4. One can therefore easily turn a discrete-time transfer function form of output-based ADRC into an error-based one by omitting the prefilter. The resulting "textbook" control loop shown in Fig. 8.11 is a structure often used as a starting point in practical digital control engineering. It should therefore also be understood as an invitation to customize this structure to a particular application's needs. Most notably a user-defined reference signal prefilter could be added to shape the control loop dynamics in a true two-degrees-of-freedom sense, as already discussed in Sect. 6.4.

8.4.3 Dual-Feedback Transfer Function Form

An equivalent of the dual-feedback transfer function form of discrete-time ADRC can also be derived for error-based ADRC. Beginning once more with the z-transform of the observer (8.43),

$$\hat{x}(z) = z^{-1} \cdot A_{ESO} \cdot \hat{x}(z) - z^{-1} \cdot b_{ESO} \cdot u_{\lim}(z) + l \cdot e(z),$$

we solve for $\hat{x}(z)$ to obtain the expression

$$\hat{x}(z) = \left(I - z^{-1} A_{ESO}\right)^{-1} \cdot \left(-z^{-1} \cdot b_{ESO} \cdot u_{\lim}(z) + l \cdot e(z)\right),$$

which can be put into the z-transform of the error-based controller (8.42),

$$u(z) = \frac{1}{b_0} \cdot (\mathbf{k}^T\ 1) \cdot \hat{\mathbf{x}}(z),$$

yielding the following control law:

$$\begin{aligned}
u(z) &= \frac{1}{b_0} \cdot (\mathbf{k}^T\ 1) \cdot \left(\mathbf{I} - z^{-1}\mathbf{A}_{\text{ESO}}\right)^{-1} \cdot \left(-z^{-1} \cdot \mathbf{b}_{\text{ESO}} \cdot u_{\lim}(z) + \mathbf{l} \cdot e(z)\right) \\
&= \underbrace{\left(-\frac{z^{-1}}{b_0} \cdot (\mathbf{k}^T\ 1) \cdot \left(\mathbf{I} - z^{-1}\mathbf{A}_{\text{ESO}}\right)^{-1} \cdot \mathbf{b}_{\text{ESO}}\right)}_{C_{\text{FBu}}(z)} \cdot u_{\lim}(z) \\
&\quad + \underbrace{\left(\frac{1}{b_0} \cdot (\mathbf{k}^T\ 1) \cdot \left(\mathbf{I} - z^{-1}\mathbf{A}_{\text{ESO}}\right)^{-1} \cdot \mathbf{l}\right)}_{C_{\text{FBy}}(z)} \cdot e(z).
\end{aligned} \qquad (8.48)$$

This means that we can abbreviate the control law (8.48) using the same two transfer functions $C_{\text{FBy}}(z)$ and $C_{\text{FBu}}(z)$ from (8.37) of the output-based counterpart:

$$u(z) = C_{\text{FBy}}(z) \cdot e(z) + C_{\text{FBu}}(z) \cdot u_{\lim}(z) \qquad (8.49)$$

$$\text{with} \quad C_{\text{FBy}}(z) = \frac{\sum_{i=0}^{N} \beta_i z^{-i}}{1 + \sum_{i=1}^{N+1} \alpha_i z^{-i}} \quad \text{and} \quad C_{\text{FBu}}(z) = z^{-1} \cdot \frac{\sum_{i=0}^{N} \gamma_i z^{-i}}{1 + \sum_{i=1}^{N+1} \alpha_i z^{-i}}.$$

The visualization of (8.49) in Fig. 8.12 allows for an easy comparison with the output-based equivalent in Fig. 8.8. To clearly indicate that the exact same transfer functions of the output-based variant are being used, we keep using the notation $C_{\text{FBy}}(z)$ with subscript "y" although now the control error e is being fed into this transfer function.

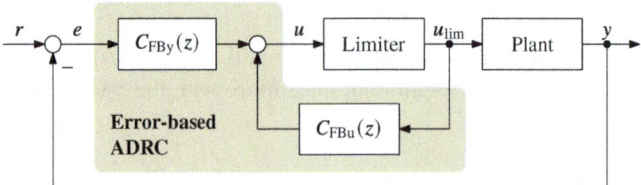

Fig. 8.12 Control loop with a dual-feedback transfer function representation of discrete-time error-based ADRC

8.4 Discrete-Time Error-Based ADRC

An efficient implementation can be found using the same ideas from Sect. 8.3. The control law (8.49) can be written making use of the common denominator of $C_{\mathrm{FBy}}(z)$ and $C_{\mathrm{FBu}}(z)$:

$$u(z) = \frac{\left(\sum_{i=0}^{N} \beta_i z^{-i}\right) \cdot e(z) + z^{-1} \cdot \left(\sum_{i=0}^{N} \gamma_i z^{-i}\right) \cdot u_{\mathrm{lim}}(z)}{1 + \sum_{i=1}^{N+1} \alpha_i z^{-i}}. \tag{8.50}$$

Using the (transposed) direct form II for (8.50) will then reduce the number of required storage variables (delays) during runtime to a minimum, as shown in Fig. 8.13 for the first- and second-order cases.

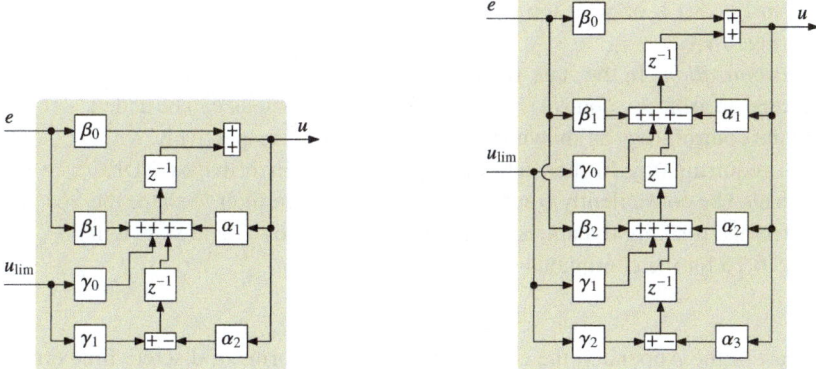

Fig. 8.13 Minimum-footprint form of the dual-feedback transfer function representation for first- and second-order discrete-time error-based ADRCs

As all coefficients α, β, γ are unchanged, they can be obtained from the tables for dual-feedback output-based ADRC in Appendix A.7. Let us now conclude our section on the three discrete-time variants of error-based ADRC with one common "cooking recipe."

> **Cooking Recipe (Discrete-Time Error-Based ADRC)**
>
> To implement and tune discrete-time error-based ADRC in one of the (a) state-space, (b) transfer function, or (c) dual-feedback transfer function forms, the following steps have to be carried out:
>
> 1. *Plant modeling:* Identify the order N and the gain parameter b_0 of the plant model.

(continued)

2. *Controller and observer tuning:* For *bandwidth parameterization*, the tuning parameters are ω_{CL} and k_{ESO}. Together with the sampling interval T_s and the plant gain b_0, these values have to be put into the respective equations for the coefficients needed in the particular form:
   - Table A.5.1 (controller and observer gains) and Table A.5.2 (observer matrix/vector elements) for the state-space form
   - Tables A.6.2 or A.6.3 for the transfer functions listed in Table A.6.1
   - Tables A.7.2 or A.7.3 for the dual-feedback transfer functions listed in Table A.7.1
3. *Implementation:*
   - State-space form: control law (8.42) with observer (8.43), as shown in Fig. 8.10
   - Transfer function form: control law (8.47), as shown in Fig. 8.11
   - Dual-feedback transfer function form: control law (8.49) as shown in Fig. 8.12, or with the efficient implementation (8.50) shown in Fig. 8.13
4. *Customization:* In the state-space and dual-feedback transfer function forms, an arbitrary control signal limitation add-on block can be added to the control loop, as shown in Figs. 8.10 and 8.12, respectively, without requiring any further anti-windup measures. Error-based ADRC can also be conveniently combined with a custom prefilter to shape the reference tracking dynamics, as described in Sect. 6.4.2 and shown in Fig. 6.16 for the continuous-time case.

As part of the Appendix, the equations for all three forms of discrete-time error-based ADRC are once more summarized in Appendix A.8.

8.5 Summary and Outlook

In this chapter we have covered three forms of discrete-time ADRC, all of which can be developed for both output-based and error-based ADRCs—see Table 8.1. Each approach comes with a different set of features and properties:

1. *State space:* A form obtained using zero-order hold discretization and the *current observer* approach. This is the discrete-time counterpart of continuous-time linear ADRC as introduced in Chap. 3.
2. *Transfer functions:* Based on z-transform of the state-space equations, discrete-time transfer functions relating the reference signal and the measured plant output to the updated control signal are developed. In doing so, the inner feedback path of the controller output is removed. While being less flexible regarding anti-windup measures, simple forms of windup protection can be implemented nevertheless. An accumulator was factored out of the feedback controller transfer

8.5 Summary and Outlook

function with this particular goal in mind. This is the discrete-time counterpart of the continuous-time transfer function representation developed in Sect. 4.2 and represents a structure often found in control engineering practice.

3. *Dual-feedback transfer functions:* Maintaining the feedback path for the (usually limited) previous controller output, this variant, which is based on two transfer functions in the feedback path, replicates the exact behavior of the state-space form. At the same time, it is possible to achieve the lowest computational footprint of all three approaches.

Table 8.1 Discrete-time forms of output-based and error-based ADRC introduced in this chapter, and where to find them

Form	Output-based	Error-based	Properties
State space	Sect. 8.1	Sect. 8.4.1	*Recognition factor:* corresponds to the original structure of continuous-time ADRC. *Control signal limitation:* flexible, built-in windup protection. *Runtime footprint:* minimum number of storage variables, but maximum number of coefficients and operations.
Transfer function	Sect. 8.2	Sect. 8.4.2	*Recognition factor:* equivalence to classical digital control with feedback controller and prefilter. *Control signal limitation:* only via clamped integrator. *Runtime footprint:* maximum number of storage variables, reduced number of coefficients and algebraic operations.
Dual-feedback TF	Sect. 8.3	Sect. 8.4.3	*Recognition factor:* fully replicates the state-space using transfer functions, but with an uncommon structure. *Control signal limitation:* flexible, built-in windup protection. *Runtime footprint:* minimum number of storage variables, coefficients, and algebraic operations.

For implementations with high sampling frequencies, for example, in power electronics applications, the runtime costs associated with a particular controller variant may not only become noticeable but even a bottleneck. In Table 8.2 we therefore compare the computational footprint of discrete-time output-based ADRC variants.

Error-based ADRC variants should be considered if either emphasis is put on the tracking performance, or a true two-degrees-of-freedom design shall be achieved using a custom prefilter in the reference signal path. In output-based ADRC, both tracking and disturbance rejection dynamics are coupled. It offers a good out-of-the-box performance for most applications with fixed, piecewise constant, or slowly changing set points, as it is tuned for a critically damped control loop response.

Table 8.2 The three discrete-time choices developed in this chapter come with different footprints. Computational costs to be expected at runtime (multiplications, additions, and storage variables; excluding costs of an optional limiter) are given for output-based ADRC of orders 1, 2, and N. For a fair comparison, direct form II implementations were taken as a basis for both transfer function approaches. Best values are highlighted. The dual-feedback variant combines favorable properties regarding both the number of operations and storage variables

Order	State space	Transfer functions	Dual-feedback TF
1st	11 mul. 10 add. **2** var.	**7** mul. **6** add. 4 var.	**7** mul. **6** add. **2** var.
2nd	19 mul. 18 add. **3** var.	11 mul. 10 add. 6 var.	**10** mul. **9** add. **3** var.
Nth	$(N^2 + 5N + 5)$ mul. $(N^2 + 5N + 4)$ add. $(N + 1)$ var.	$(4N + 3)$ mul. $(4N + 2)$ add. $(2N + 2)$ var.	$(\mathbf{3N + 4})$ mul. $(\mathbf{3N + 3})$ add. $(N + 1)$ var.

Next Steps

In this chapter, we have selected and derived several basic discrete-time ADRC variants. The next three chapters build on those results showing various aspects related to the implementation of ADRC in practical scenarios.

- Chapter 9 describes various features that can further facilitate a successful deployment of ADRC thus contributing to closing the theory-practice gap.
- Chapter 10 supports the transition from equations to actual software implementation, both in model-based and handwritten source code implementations.
- Chapter 11 documents the use of the above discrete-time ADRC in actual application examples, showing how the manufactured "cooking recipes" perform when applied to real control problems, which call for actual hardware implementations.

References

1. Franklin, G.F., Powell, D., Workman, M.L.: Digital Control of Dynamic Systems, 3rd edn. Addison-Wesley Longman Publishing, Boston, MA, USA (1997)
2. Miklosovic, R., Radke, A., Gao, Z.: Discrete implementation and generalization of the extended state observer. In: Proceedings of the American Control Conference, pp. 2209–2214 (2006). https://doi.org/10.1109/ACC.2006.1656547
3. Herbst, G., Madonski, R.: Tuning and implementation variants of discrete-time ADRC. Control Theory Technol. **21**, 72–88 (2023). https://doi.org/10.1007/s11768-023-00127-0
4. Herbst, G.: Transfer function analysis and implementation of active disturbance rejection control. Control Theory Technol. **19**, 19–34 (2021). https://doi.org/10.1007/s11768-021-00031-5

References

5. Peng, Y., Vrančič, D., Hanus, R.: Anti-windup, bumpless, and conditioned transfer techniques for PID controllers. IEEE Control. Syst. Mag. **16**(4), 48–57 (1996). https://doi.org/10.1109/37.526915
6. Åström, K.J., Hägglund, T.: Advanced PID Control. International Society of Automation, New York (2006)
7. Herbst, G.: A minimum-footprint implementation of discrete-time ADRC. In: Procedings of the European Control Conference, pp. 107–112 (2021). https://doi.org/10.23919/ECC54610.2021.9655120
8. Oppenheim, A., Schafer, R.: Discrete-Time Signal Processing, 3rd edn. Prentice Hall, New York (2010)

Open Access This chapter is licensed under the terms of the Creative Commons Attribution 4.0 International License (http://creativecommons.org/licenses/by/4.0/), which permits use, sharing, adaptation, distribution and reproduction in any medium or format, as long as you give appropriate credit to the original author(s) and the source, provide a link to the Creative Commons license and indicate if changes were made. The images or other third party material in this chapter are included in the chapter's Creative Commons license, unless indicated otherwise in a credit line to the material. If material is not included in the chapter's Creative Commons license and your intended use is not permitted by statutory regulation or exceeds the permitted use, you will need to obtain permission directly from the copyright holder.

Chapter 9
Practical Aspects

Abstract Tuning a controller is not always the hardest part. Its implementation in a real-world control loop requires additional efforts: One might have to be able to (gracefully) turn the controller on or off, modify its parameters while in operation, and deal with unwanted behaviors such as measurement noise, actuator limitations, or time delays in the control loop. This is usually easier said than done. In this chapter, we examine some of these issues and try to offer solutions that maintain the simplicity we consider a vital part of the "ADRC spirit."

9.1 Controller Output Limitation

9.1.1 Magnitude Limitation and Windup Protection

In every practical control loop, the manipulated variable—fed by the controller output—is subject to certain limitations. This leads to problems if the controller wants to drive the actuator beyond its limits while trying to reach or maintain a set point: As long as a control error persists, the controller tries to cure the issue by applying more of the same medicine, leading to an ever-increasing controller output magnitude. This phenomenon is often called *integrator windup*,[1] since the controller accumulates nonzero control errors due to its integral behavior.

There is a multitude of model-free and model-based approaches to prevent windup [2]. In practical (often, PID) controllers, simple ("ad hoc") solutions are usually employed, such as conditional integration or back-calculation [3]. However, each variant comes with a different set of properties and some with additional parameters to tune. This puts an extra burden on the designer of a control loop.

ADRC can be affected by windup, as well, since estimating the *total disturbance* creates an integral term, which we have seen in the transfer function representations.

---

[1] Due to its practical relevance, we are only covering windup occurring within the controller in this book—so-called *controller windup*, as opposed to *plant windup* [1].

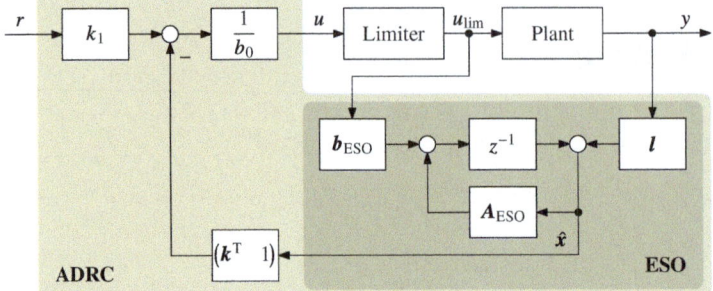

Fig. 9.1 Discrete-time ADRC, state-space implementation, enhanced by a control signal limitation block. Feeding back the limited controller output u_{lim} instead of u (as in Fig. 8.4) is the only measure required to prevent windup in ADRC

And now for the good news, since ADRC is built around an observer, there is one simple yet effective solution that does not require additional tuning, making use of the so-called *observer technique* [1]. As we have already hinted at several times, especially in Chap. 8, feeding the limited control signal back to the observer prevents windup [4]. Figure 9.1 shows the resulting control loop structure.

To achieve magnitude limitation of the control signal—which, in our opinion, should be part of every practical controller—the additional "Limiter" block consists of a hard upper/lower limit for u:

$$u_{\text{lim}} = \text{sat}_{u_{\min}}^{u_{\max}}(u) = \begin{cases} u_{\min} & , u < u_{\min} \\ u & , u_{\min} \leq u \leq u_{\max} \\ u_{\max} & , u > u_{\max}. \end{cases} \quad (9.1)$$

To prevent windup, the observer is now being fed by the limited signal u_{lim} instead of the unlimited controller output u—it is as simple as that. Do you believe that? We can find out using the dual-feedback transfer function representation (8.37) from Sect. 8.4.3, where we derived two discrete-time transfer functions from the measured plant output y and the limited controller output u_{lim} back to the updated (unlimited) controller output u. If we put in the α, β, γ coefficients provided in Appendix A.7 for *bandwidth parameterization*, we can obtain the DC gain of these transfer functions by evaluating at zero frequency (i.e., $z = 1$), yielding

$$C_{\text{FB}u}(z=1) = 1 \quad \text{and} \quad C_{\text{FB}y}(z=1) = \frac{1 - z_{\text{CL}}}{b_0 T_s} \quad \text{for} \quad N = 1, \quad \text{and}$$

$$C_{\text{FB}u}(z=1) = 1 \quad \text{and} \quad C_{\text{FB}y}(z=1) = \frac{(1 - z_{\text{CL}})^2}{b_0 T_s^2} \quad \text{for} \quad N = 2.$$

Evidently both $C_{\text{FB}u}$ and $C_{\text{FB}y}$ are of proportional type. The controller output u will therefore not wind up if both u_{lim} and y are stuck at certain values but converge to a finite value regardless of the control error.

9.1 Controller Output Limitation

In Fig. 9.2, ADRC-based control loops building on our go-to example from Sect. 5.2 are being compared, with and without this simple anti-windup technique. It can be seen that, with anti-windup, overshoot of y and build-up of u are prevented, and u recovers nicely from being in saturation.

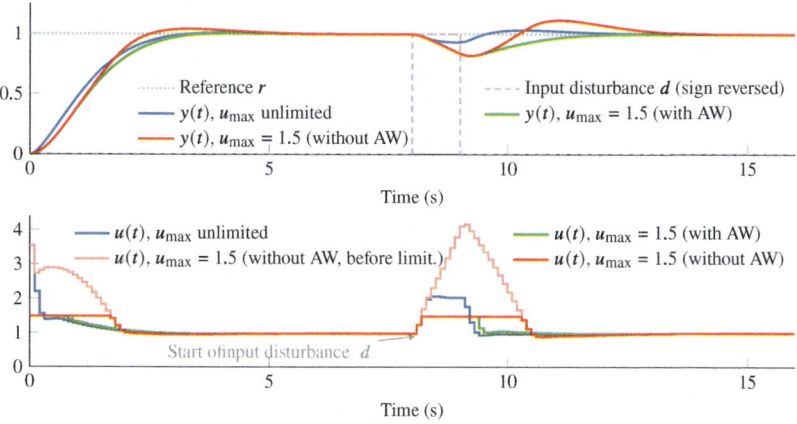

Fig. 9.2 ADRC with magnitude limitation: closed-loop simulation of the example in Fig. 5.4, this time with a discrete-time state-space implementation of ADRC with $T_s = 0.1$ s. Tuning and plant parameters are as described in Sect. 5.2; only the bandwidth was increased to $\omega_{CL} = 2\pi \cdot 0.3$ rad/s. Three variants are compared: unlimited controller output (blue), an externally applied limitation (without feedback to ADRC, i.e., without anti-windup—the control signal is shown both before and after the limitation block here, red colors), and the scheme of Fig. 9.1 (with anti-windup via feedback of $u_{\lim}$ to the observer, green color). Using reference and disturbance steps, the controllers are deliberately driven into saturation. The simple anti-windup scheme of Fig. 9.1 prevents overshoot and windup of the integrator state within ADRC

9.1.2 Rate Limitation

On top of magnitude limitation, more advanced restrictions to the control signal are, within certain limits, possible—which is the reason we put such a generic "Limiter" block in Fig. 9.1. One notable and practically relevant example is the inclusion of an additional rate limitation. During transients or after sudden load changes in a control loop, this enforces a maximum possible velocity for control signal changes, which can also help to be easier on the actuator. A discrete-time implementation for both magnitude and rate limitation is shown in Fig. 9.3 and governed by

$$u_{\lim}(k) = \mathrm{sat}_{u_{\min}}^{u_{\max}}\left(u_{\lim}(k-1) + \mathrm{sat}_{\Delta u_{\min} \cdot T_s}^{\Delta u_{\max} \cdot T_s}\left(u(k) - u_{\lim}(k-1)\right)\right). \qquad (9.2)$$

Fig. 9.3 Discrete-time limitation block adapted from [4]: Using two (symmetric or asymmetric) saturation elements, the output signal u_{lim} can be limited in both magnitude and rate, cf. (9.2)

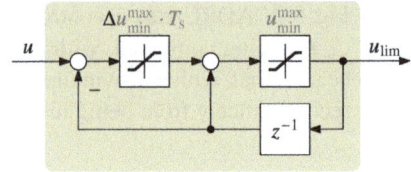

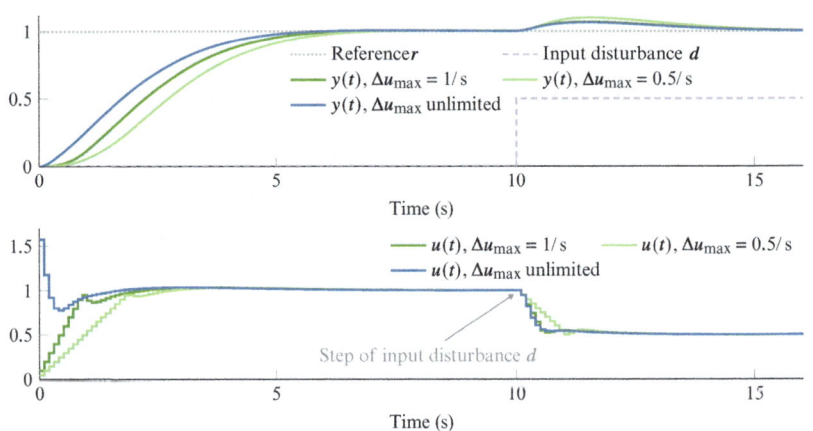

Fig. 9.4 ADRC with rate-limited controller output: closed-loop simulation of the example in Fig. 5.4, with a discrete-time state-space implementation of ADRC with $T_s = 0.1$ s. Tuning and plant parameters are as described in Sect. 5.2. In this example, there are no magnitude limits set for the controller output u, but a rate limitation is in place for two variants (green colors). If chosen to be strict (slow), a rate limitation will reduce the controller's ability to reject disturbances and follow the reference signal with the initially designed bandwidth

The simulation example in Fig. 9.4 compares controllers with two different rate limits to an unlimited controller. Again, $u(k)$ recovers nicely from being rate-limited, which underlines the effectiveness of the observer-based anti-windup technique. The price to pay for a slowed-down control signal might be a reduced ability to quickly compensate disturbances or to achieve the desired closed-loop response.

One must always keep in mind that the anti-windup extension of ADRC discussed here belongs to the category of simple, "ad hoc" techniques. As such, it cannot do wonders and compete with advanced model-based approaches such as MPC. It is very possible to provoke oscillations by choosing the rate limit too strict—which, on the other hand, is not advisable due to the decreased responsiveness of the controller anyway. A rate limit should be carefully selected and, if possible, evaluated using a closed-loop simulation. Having a rate limit as an easy-to-implement option for ADRC in the first place, however, is a very welcome addition and a selling point compared to traditional PID-type controllers.

9.2 Bumpless Transfer

One of the aspects that practical controllers and their "textbook" siblings differ in might appear trivial, despite being of great importance: the ability of smoothly *enabling* a controller, i.e., without inducing unnecessary control actions. This has never appeared to be a problem in any of the simulation examples presented so far, since all of them were being simulated in closed-loop mode starting from zero initial conditions for both plant and controller. Nonzero initial conditions can, for example, be caused in practice by the plant's ambient conditions—think of the room temperature in a temperature control task. The desired way of transitioning to closed-loop control from manual control mode, or switching between different controllers, is often called *bumpless transfer*, as no "bumps" shall occur in either the controller output or the controlled variable. Similar issues can occur when retuning a controller while in operation, where a sudden parameter change may lead to unwanted drastic changes in the controller output. In this section, we want to address this class of problems.

9.2.1 Transfer from Manual to Automatic Control

To demonstrate what can go wrong without proper initialization, let us pick up the example from Sect. 5.2 once more. In the first set of results shown in Fig. 9.5, it is controlled by discrete-time ADRC that is enabled after being in a settled, nonzero state due to manual control. Enabling ADRC with zero internal state in this scenario leads to terrible results—no signs of a "bumpless" transfer, we must do better!

We consider two discrete-time ADRC forms especially relevant for practical use: the state-space and the dual-feedback transfer function implementation, as they are functionally equivalent also regarding their anti-windup possibilities. Initialization procedures to achieve bumpless transfer for these two are therefore being discussed in the following. If the reference signal value is chosen to match $r = y$ and the plant being in a settled state before, the transition to closed-loop control should then be bumpless, i.e., not be noticeable in both the control signal and the plant output.

- *State-space implementation:* As outlined in [4], one option involves enabling the observer earlier than the controller and feeding the externally provided control signal u and plant output y to the observer. This approach can be called *tracking*, and if this is done in time to allow the observer to reach a settled state before the controller output is being activated, bumpless transfer can be achieved. If the observer state is not yet fully settled, the transfer will still be (slightly) bumpy, as specifically demonstrated in Fig. 9.5 (second example).

  Another option allows to simultaneously enable controller and observer at time step k by means of *direct initialization* of the observer state to a meaningful previous value $\hat{x}(k-1)$. Given a measurement value of the plant output $y(k-1)$ and the initial control action $u(k-1)$ (e.g., from manual control), a simple

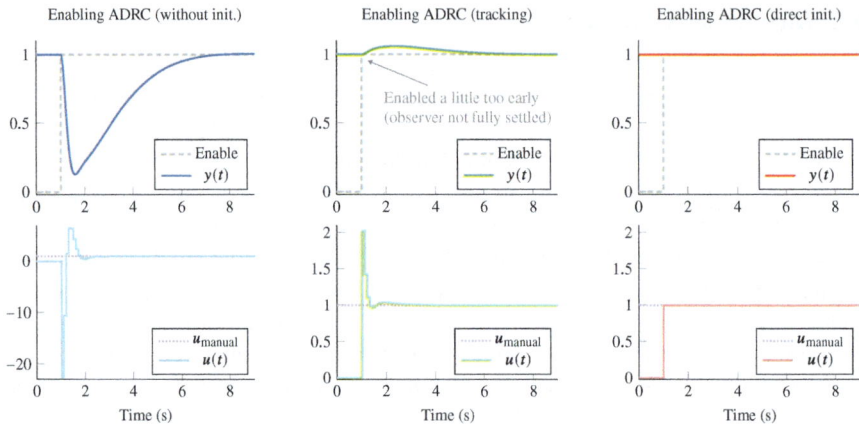

Fig. 9.5 Bumpy and bumpless transfer to automatic control: enabling discrete-time state-space ADRC in a nonzero initial state of the plant $P(s) = \frac{1}{s^2+2s+1}$ from Sect. 5.2. The transition between manual control with $u_{\text{manual}} = 1$ and closed-loop control with $r = y = 1$ occurs at time $t = 1$ s. Three variants are compared: (left, blue colors) no initialization, i.e., all internal state variables are zero when switching on, (center, green colors) tracking of y and u_{manual} by the observer before switching on the controller output—here purposefully a little too early, and (right, red colors) direct initialization of the observer state prior to enabling the controller output. Obviously, enabling ADRC without initialization—i.e., with zero-valued internal state—can be a bad idea, leading to enormous control action spikes. Enabling the observer well before the controller to let the ESO track the manual control action and current plant output can resolve this issue if the controller does not need to be enabled instantly. Direct initialization allows to bumplessly switch on both observer and controller at the same time

initialization strategy consists of setting the estimated output to the measured value, i.e., $\hat{x}_1(k-1) = \hat{y}(k-1) = y(k-1)$, and adjusting the *total disturbance* such that the controller output will match the given value $u(k-1)$ if $r = y$ is being chosen:

$$\hat{x}(k-1) = \begin{pmatrix} y(k-1) \\ \mathbf{0}^{(N-1)\times 1} \\ -b_0 \cdot u(k-1) \end{pmatrix}. \tag{9.3}$$

Attention! It is important to note that the z^{-1} element in the observer implementation of Fig. 9.1 does not hold the state $\hat{x}$, since $\hat{x}(k)$ is being formed by the output of said z^{-1} element plus $\mathbf{l} \cdot y(k)$. To initialize the observer output to the desired value given in (9.3), one must therefore set the $(N + 1)$ storage variables of the z^{-1} element to the value $\hat{x}(k-1) - \mathbf{l} \cdot y(k)$. The third example shown in Fig. 9.5 uses this approach to achieve bumpless transfer.

- *Dual-feedback transfer function implementation:* For the minimum-footprint variants shown in Fig. 8.9, we can adapt the initialization steps presented in [5] to the form introduced in this book.

9.2 Bumpless Transfer

1. Denote the sum of the two feedback transfer function outputs as the intermediate variable f, and compute its value from the recent measurement $y(k-1)$ and the externally provided previous control action $u(k-1)$:
$$f(k-1) = u(k-1) - \frac{k_1}{b_0} \cdot y(k-1).$$
2. Initialize the storage variable x_{N+1} belonging to the lowermost unit delay:
$$x_{N+1}(k-1) = -\alpha_{N+1} \cdot f(k-1) + \gamma_N \cdot u(k-1).$$
3. Recursively initialize the upper N storage variables x_i in reverse order, i.e., $\forall i = N, N-1, \ldots, 1$:
$$x_i(k-1) = x_{i+1}(k-1) - \alpha_i \cdot f(k-1) - \beta_i \cdot y(k-1) + \gamma_{i-1} \cdot u(k-1).$$

For error-based ADRC, things are a bit easier since the estimated variables are derived from the control error—which is zero when transitioning from a settled state to closed-loop control with $r = y$. Nonetheless, initialization is necessary to provide the desired output value upon enabling the controller.

- *Error-based ADRC, state-space implementation:* For direct initialization of the observer state, (9.3) gets simplified under these conditions and now reads
$$\hat{x}(k-1) = \begin{pmatrix} \mathbf{0}^{N \times 1} \\ b_0 \cdot u(k-1) \end{pmatrix}. \tag{9.4}$$

In a settled state $r = y$, which implies $e = r - y = 0$, this value can directly be written into the $(N+1)$ storage variables of the z^{-1} element of the observer in Fig. 8.10. Alternatively, the tracking approach can, of course, also be adapted to the initialization of error-based ADRC.

- *Error-based ADRC, dual-feedback transfer function implementation:* Similar to the steps described above for output-based ADRC, initializing the minimum-footprint variants shown in Fig. 8.13 can be done for the state variables (unit delays) in a recursive manner. Given an externally provided previous control action $u(k-1)$ and assuming zero control error for the time instant of enabling the controller, one obtains the following procedure.

1. Initialize the storage variable x_{N+1} belonging to the lowermost unit delay:
$$x_{N+1}(k-1) = (\gamma_N - \alpha_{N+1}) \cdot u(k-1).$$
2. Recursively initialize the upper N storage variables x_i in reverse order, i.e., $\forall i = N, N-1, \ldots, 1$:
$$x_i(k-1) = x_{i+1}(k-1) + (\gamma_{i-1} - \alpha_i) \cdot u(k-1).$$

For x_1, this initialization necessarily leads to the value $x_1(k-1) = u(k-1)$.

Exactly the same procedures can also be used when switching from a different controller to ADRC. Finally, if present, an output limitation block must be initialized in all variants before enabling the controller output such that its initial output matches the given previous control action $u(k-1)$.

9.2.2 Bumpless Parameter Changes

Another desirable feature of a real-world controller is the ability to modify its parameters while being in operation. Remembering the ingredients of ADRC first summarized in Sect. 3.1.5, we can group parameter modifications in three categories: changes to the *critical gain parameter* b_0, changes to the state-feedback controller, and changes to the observer. Note that, when using *bandwidth parameterization* with ω_{CL} and k_{ESO}, a modification of ω_{CL} would affect controller and observer at the same time. From the factor $\frac{1}{b_0}$ in the control law (3.15), it becomes obvious that a sudden change of b_0 would immediately result in an abrupt change of the controller output—an undesirable behavior.

Avoiding jumps of the controller output when modifying its parameters is a problem very much related to that of bumpless transfer: One can think of it as switching from one controller to another, with a different parameter set. At least in a settled state $r = y$, the control signal u should remain constant when doing so. It, therefore, suggests itself to reuse the solutions discussed before for the bumpless transfer problem. The procedure for a parameter modification would then be:

1. Compute the controller output with the previous parameter set.
2. Apply the parameter changes, i.e., if one of the input parameters to ADRC tuning (such as b_0, ω_{CL}, or k_{ESO}) is modified, update all dependent matrix or transfer function coefficients for the implementation chosen.
3. Reinitialize the internal controller/observer state variables such that the controller will maintain the previously computed output value, e.g., using (9.3).

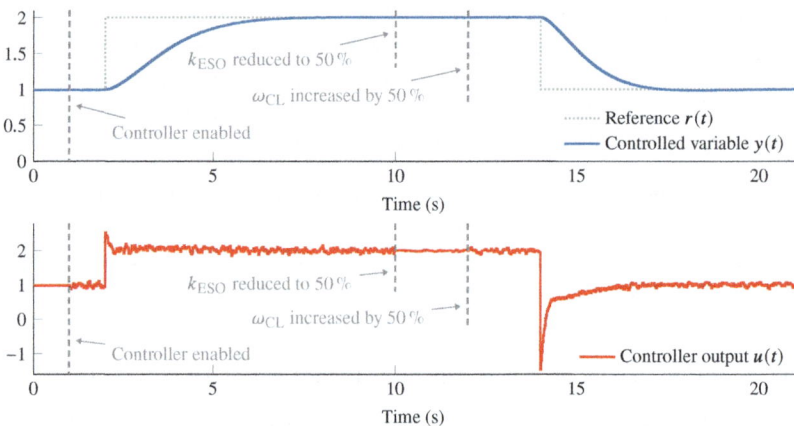

Fig. 9.6 Bumpless transfer and bumpless parameter changes for discrete-time ADRC with $N = 2$ and $T_s = 20$ ms. Initial tuning parameters are $\omega_{\text{CL}} = 2\pi \cdot 0.2$ rad/s, $k_{\text{ESO}} = 10$, and $b_0 = 1$. Normalized plant $P(s) = \frac{1}{s^2 + 2s + 1}$ from Sect. 5.2. Measurement noise is added with $\sigma = 0.002$. As desired, the bandwidth changes do not lead to bumps in either y or u but can be recognized from the varying impact of measurement noise on u, as well as the settling time

9.3 Dealing with Measurement Noise

A simulation example of implementing bumpless parameter changes using this approach is presented in Fig. 9.6. As a final note, it might be argued if bumps in u should be "allowed" when switching controllers or modifying parameters in the midst of a transient. In the eyes of the controller taking over, starting from an initial condition $r \neq y$ could be interpreted as the beginning of a step response, which usually involves a jump in u. Steering clear of this discussion, we limited ourselves in this book to providing solutions for bumpless transfer and parameter changes when the plant is in a settled state $r = y$.

9.3 Dealing with Measurement Noise

Measurement noise is almost inevitable in any practical control loop, although the severity of the problem will, of course, depend on the considered application. One result of the *gang of six* analysis performed in Sect. 5.3.3 was that high-frequency measurement noise will impact the controller output depending on the amount of damping provided in the transfer function G_{un} in that frequency range—which, in turn, depends on damping provided by the feedback controller transfer function C_{FB}.

In Fig. 5.8 of Sect. 5.3.3, we have already seen that the choice of observer bandwidth affects G_{un} and will require a compromise. Larger values of k_{ESO} would—in theory—be "nice to have" regarding disturbance rejection and maintaining closed-loop dynamics even under plant uncertainties. On the other hand, some upper limit to k_{ESO} (or, more general, to the observer speed) will be imposed by the maximum amount of noise admissible in the controller output.

What we have not seen yet, however, is the fact that all of this becomes even more of a problem for increased plant orders N. In Fig. 9.7 we demonstrate this fact with two examples controlling normalized first- and second-order plants with ADRC implemented in discrete-time form. The same levels of measurement noise are inducing vastly different levels of noise in the control signal, and this will worsen fast if the order is increased beyond $N = 2$.

So what can be done to mitigate the impact of measurement noise? First, we want to briefly discuss two options to reduce the noise sensitivity without requiring structural modifications or additions to the controller or control loop.

- *Bandwidth selection:* An obvious approach to this problem is to reduce the bandwidth when tuning ADRC, particularly regarding the observer dynamics. It was evident from Fig. 5.8 in Sect. 5.3.3 that the choice of k_{ESO} quite drastically affects the sensitivity of the controller output to high-frequency measurement noise. To demonstrate this fact with a time-domain example, Fig. 9.8 builds on the second-order example from Fig. 9.7 and compares observer designs with $k_{ESO} = 5$ and $k_{ESO} = 10$. Solving the noise sensitivity problem with reduced k_{ESO} values, on the other hand, must be traded in for reduced disturbance rejection abilities and the fact that closed-loop dynamics will increasingly fall short of the design goals, as visible from closed-loop step responses in Fig. 9.7, as well.

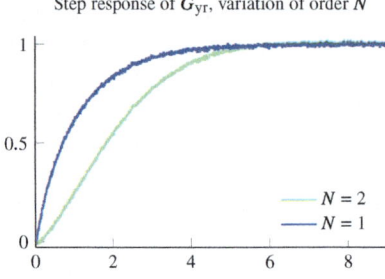 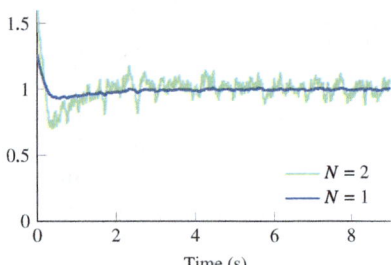

Fig. 9.7 Influence of the system order N on the sensitivity of the controller output to Gaussian measurement noise with $\sigma = 0.005$. The normalized first-order plant is $P_1(s) = \frac{1}{s+1}$ and the second-order plant $P_2(s) = \frac{1}{s^2+2s+1}$. Both control loops are being tuned for the same bandwidth $\omega_{\mathrm{CL}} = 2\pi \cdot 0.2\,\mathrm{rad/s}$, $k_{\mathrm{ESO}} = 5$, and with $b_0 = 1$. Sampling interval of the discrete-time controller implementation is $T_s = 10\,\mathrm{ms}$. With increasing system (= controller) order N, the impact of measurement noise on the controller output becomes drastically larger

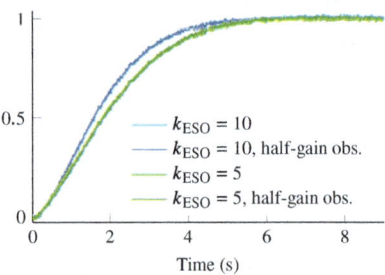 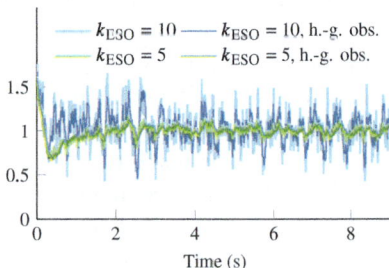

Fig. 9.8 Influence of tuning parameters on the sensitivity of the controller output to Gaussian measurement noise with $\sigma = 0.005$ in a control loop with a normalized second-order plant $P(s) = \frac{1}{s^2+2s+1}$. Other ADRC parameters remain fixed at $N = 2$, $b_0 = 1$, and $\omega_{\mathrm{CL}} = 2\pi \cdot 0.2\,\mathrm{rad/s}$. Sampling interval of the discrete-time controller implementation is $T_s = 10\,\mathrm{ms}$. For each of the observer bandwidth choices ($k_{\mathrm{ESO}} = 5$, green colors; $k_{\mathrm{ESO}} = 10$, blue colors), a half-gain tuning variant was added (thinner, darker lines), demonstrating the ability to reduce the noise sensitivity while maintaining the dynamics of the controlled variable

Table 9.1 Mean absolute deviation (MAD) and root mean square (RMS) values of the noise levels in the examples of Fig. 9.8, computed for $k_{\mathrm{ESO}} = 5$ and $k_{\mathrm{ESO}} = 10$, both without and with *half-gain tuning* for the observer (cases labeled "HGO"). These values were obtained from steady-state simulations, averaging 100000 samples. Caused by Gaussian noise, the MAD and RMS levels are related by a factor of $\sqrt{2/\pi}$ [6]. Using *half-gain tuning* for the observer reduces the noise levels by approximately more than 25 %, without noticeable impact on the closed-loop dynamics

Criterion	$k_{\mathrm{ESO}} = 5$	$k_{\mathrm{ESO}} = 5$, HGO	$k_{\mathrm{ESO}} = 10$	$k_{\mathrm{ESO}} = 10$, HGO
MAD	0.03840	0.02761	0.15555	0.11585
RMS	0.04815	0.03462	0.19501	0.14534

- *Half-gain tuning:* A more subtle solution to tuning ADRC with reduced noise sensitivity in mind was proposed in [7], called *half-gain tuning*. It can be applied to tuning both controller and observer within ADRC, and the application of this method is very simple in continuous-time domain: Firstly, compute controller and/or observer gains using *bandwidth parameterization*; then reduce these gains to 50 % afterward. This results in closed-loop dynamics that—when applied only to the observer—are almost indistinguishable from regular *bandwidth parameterization* but lead to reduced high-frequency gains and hence improved noise suppression. In Fig. 9.8 two additional test cases demonstrate the effect of using *half-gain tuning* for the observer.

  Applying *half-gain tuning* to discrete-time ADRC involves a few more steps: Compute the continuous-time eigenvalues of the closed outer loop for the virtual plant (3.17) using halved controller gains $\boldsymbol{k}^T$ or the eigenvalues of the closed-loop observer (3.16) using halved observer gains $\boldsymbol{l}$. Map these poles to the discrete-time domain, and then perform pole placement for the discrete-time virtual plant (8.12) or the discrete-time observer (8.5), yielding the desired results for $\boldsymbol{k}^T$ or $\boldsymbol{l}$.

  To quantify the noise levels indicated only visually in the examples of Fig. 9.8, Table 9.1 gives RMS and MAD values for the controller output for these two options (k_{ESO} variation and *half-gain tuning*). If freedom to modify the controller or control loop structure is available, there are additional options to mitigate noise problems.

- *Additional low-pass filter:* As demonstrated in Sect. 5.4.2, an additional pole added to the plant might—if fast enough, of course—be barely noticeable in the closed-loop dynamics. Adding a simple low-pass filter in the feedback path with a time constant being only a fraction of the plant time constant is therefore one possible option to reduce the impact of measurement noise.
- *Enhanced ADRC structures:* ADRC can be equipped with additional countermeasures against measurement noise. As one example, cascading multiple extended state observers is proposed in [8]. This, however, comes with some drawbacks on its own, such as additional tuning parameters or the burden of selecting the correct structure [9]. It should therefore be considered an ongoing research topic to keep an eye on.

9.4 Control of Plants with Dead Time

Many practical processes exhibit time delays—for example, due to actual transport mechanisms or as part of a simplified plant modeling, where a "low-pass plus time delay" is a sufficiently accurate description of higher-order dynamics. In either scenario, models of a plant such as those given in Table 3.1 must be extended to take the time delay (dead time T_d, in seconds) into account. For a first-order plant, $P(s)$ then reads

$$P(s) = \frac{K}{Ts+1} \cdot e^{-T_d s}. \tag{9.5}$$

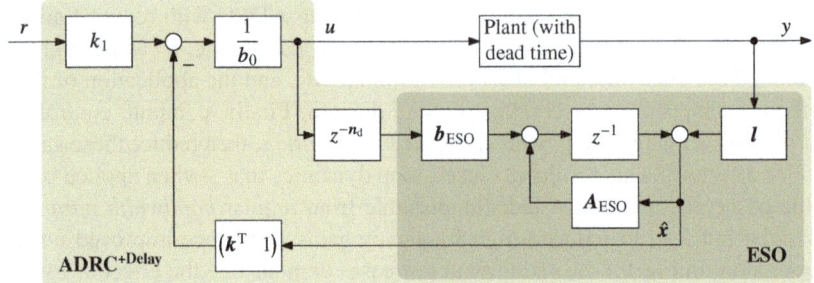

Fig. 9.9 Block diagram of a control loop with discrete-time ADRC, as in Fig. 8.4, but for a plant with dead time. The control signal entering the observer is delayed by n_d samples to synchronize the observer with the time-delayed plant

So far, we did not consider plants with time delays in this book. But if information about the plant dead time is available during control design, it appears reasonable that one can make beneficial use of that information.

For ADRC, one simple and practical solution is to *synchronize* the observer (which has a plant model inside) with the time-delayed plant [10]. In that case, the implementation of discrete-time ADRC in state space could be realized based on the control law (8.1), but with the following modification to the input signal u of the discrete-time ESO:

$$\hat{x}(k) = A_{\mathrm{ESO}} \cdot \hat{x}(k-1) + b_{\mathrm{ESO}} \cdot u(k-1-n_d) + l \cdot y(k). \tag{9.6}$$

In (9.6), terms A_{ESO}, b_{ESO}, and l are identical to the original observer in (8.5). From (9.6), it is evident that synchronization of ESO is being realized solely through an additional delay of the fed-back control signal by n_d samples. In Fig. 9.9, we show the control loop with an accordingly modified discrete-time ADRC implementation. The value of n_d must be chosen to let the discrete-time delay approximately correspond to T_d:

$$n_d \approx \frac{T_d}{T_s}. \tag{9.7}$$

To provide some intuition if and when time delays can be a problem in a control loop when *not* considering them, Fig. 9.10 presents closed-loop step responses of unmodified ADRC implementations controlling a first-order plant with increasing dead time values T_d. As visible, larger time delays induce oscillations, and the control loop can quickly become unstable beyond that.

While the simple solution of (9.6) as part of Fig. 9.9 cannot achieve wonders, a mitigation of delay-induced problems is possible. This is being demonstrated in the examples shown in Fig. 9.11, which build on the worst-case simulation of Fig. 9.10 but now synchronize the observer with the plant. Even if one does not exactly know the actual dead time, the closed-loop response is considerably improved.

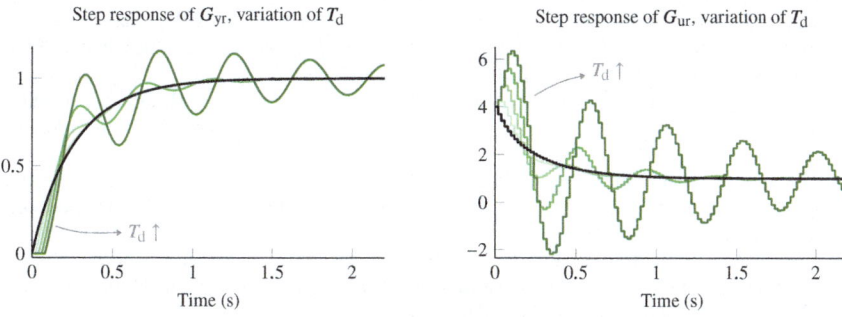

Fig. 9.10 Effect of adding dead time T_d to a normalized first-order plant $P(s) = \frac{1}{s+1}$ (nominal case: thick black line) in a control loop with first-order discrete-time ADRC ($T_s = 20$ ms). Values of T_d are ranging from 20 ms to 80 ms (increasingly dark colors). A further increase of T_d would render the control loop unstable

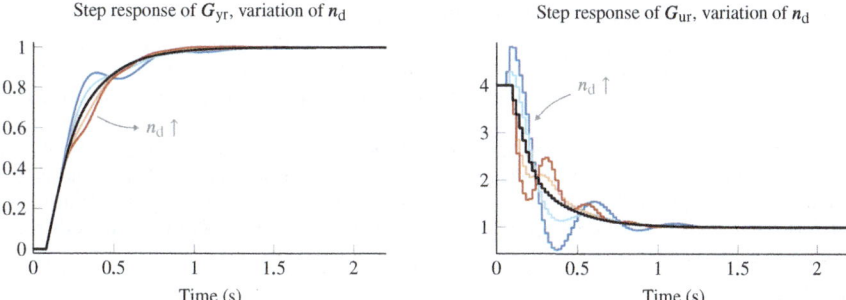

Fig. 9.11 Compensating the effect of the worst-case dead time ($T_d = 80$ ms) from Fig. 9.10 by delaying the u-input of the observer by n_d samples (with sampling interval $T_s = 20$ ms), as shown in Fig. 9.9. To demonstrate the effect of missing the actual dead time value, i.e., $n_d \neq \frac{T_d}{T_s} = 4$ (nominal case, thick black line), values of the applied delay are varying between $n_d = 2$ (underestimation, blue colors) and $n_d = 6$ (overestimation, red colors)

References

1. Hippe, P.: Windup in control: its effects and their prevention. In: Advances in Industrial Control. Springer, London (2006). https://doi.org/10.1007/1-84628-323-X
2. Tarbouriech, S., Garcia, G., ao Manoel Gomes da Silva Jr., J., Queinnec, I.: Stability and Stabilization of Linear Systems with Saturating Actuators. Springer, London (2011). https://doi.org/10.1007/978-0-85729-941-3
3. Visioli, A.: Practical PID control. In: Advances in Industrial Control. Springer, London (2006). https://doi.org/10.1007/1-84628-586-0
4. Herbst, G.: Practical active disturbance rejection control: bumpless transfer, rate limitation, and incremental algorithm. IEEE Trans. Ind. Electron. **63**(3), 1754–1762 (2016). https://doi.org/10.1109/TIE.2015.2499168
5. Herbst, G.: A minimum-footprint implementation of discrete-time ADRC. In: Proceedings of the European Control Conference, pp. 107–112 (2021). https://doi.org/10.23919/ECC54610.2021.9655120

6. Huber, P.J.: Robust Statistical Procedures, 2nd edn. In: CBMS-NSF Regional Conference Series in Applied Mathematics. Society for Industrial and Applied Mathematics, Philadelphia, PA, USA (1996). https://doi.org/10.1137/1.9781611970036
7. Herbst, G., Hempel, A.J., Göhrt, T., Streif, S.: Half-gain tuning for active disturbance rejection control. In: Proceedings of the IFAC World Congress, pp. 1319–1324 (2020). https://doi.org/10.1016/j.ifacol.2020.12.1864
8. Łakomy, K., Madonski, R.: Cascade extended state observer for active disturbance rejection control applications under measurement noise. ISA Trans. **109**, 1–10 (2021). https://doi.org/10.1016/j.isatra.2020.09.007
9. Ahmad, S., Ali, A.: Unified disturbance-estimation-based control and equivalence with IMC and PID: case study on a DC-DC boost converter. IEEE Trans. Ind. Electron. **68**(6), 5122–5132 (2021). https://doi.org/10.1109/TIE.2020.2987269
10. Herbst, G.: A simulative study on active disturbance rejection control (ADRC) as a control tool for practitioners. Electronics **2**(3), 246–279 (2013). https://doi.org/10.3390/electronics2030246

Open Access This chapter is licensed under the terms of the Creative Commons Attribution 4.0 International License (http://creativecommons.org/licenses/by/4.0/), which permits use, sharing, adaptation, distribution and reproduction in any medium or format, as long as you give appropriate credit to the original author(s) and the source, provide a link to the Creative Commons license and indicate if changes were made.

The images or other third party material in this chapter are included in the chapter's Creative Commons license, unless indicated otherwise in a credit line to the material. If material is not included in the chapter's Creative Commons license and your intended use is not permitted by statutory regulation or exceeds the permitted use, you will need to obtain permission directly from the copyright holder.

Chapter 10
Software Implementation

Abstract The "cooking recipes" in this book shall provide guidance and encouragement for custom ADRC implementations. To demonstrate that the transition from principles to practice is immediately possible based on the results from previous chapters, we will implement discrete-time variants in software form in this chapter; both using a model-based environment and using manual C language coding.

10.1 Implementation in MATLAB/Simulink

Widely used in both industry and academia, MATLAB/Simulink[1] offers a programming and simulation environment that is particularly well-suited for design, analysis, and implementation of control systems. Later in this book, in Appendix B, we will discuss existing ADRC solutions for Simulink. In this section, however, we want to demonstrate that a custom Simulink implementation of ADRC can be created straightforwardly, and encourage readers to do so on their own.

Using the block diagrams and equations given in this book, especially in the condensed compilation of ADRC variants in Appendix A, it is a straightforward task to implement ADRC on one's own in a model-based environment such as Simulink. We will demonstrate that by implementing the control loop from the second-order example in Sect. 5.2, this time with the discrete-time state-space variant of ADRC introduced in Sect. 8.1, which is summarized in condensed form in Appendix A.5.

The model-based ADRC implementation consists of:

- the discrete-time control law (A.5.1);
- the discrete-time observer (A.5.2) derived in Sect. 8.1;
- a (magnitude) limitation of the controller output signal discussed in Sect. 9.1, including the simple but effective windup protection provided by feeding the limited controller output back into the observer.

---

[1] *MATLAB* and *Simulink* are registered trademarks of *The MathWorks, Inc.*

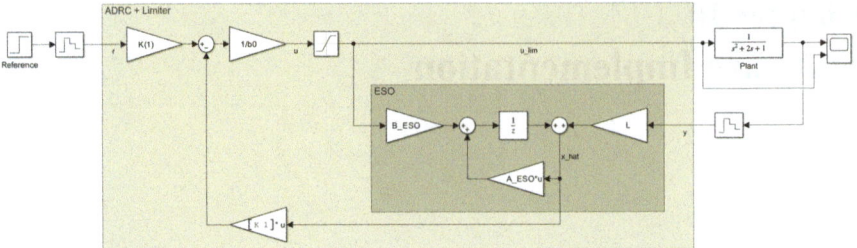

Fig. 10.1 Custom Simulink implementation of the control loop from Sect. 5.2 using discrete-time state-space ADRC as summarized in Appendix A.5. The required initialization script is given in Listing 10.1. Zero-order hold blocks and the unit delay are parameterized with the sampling time Ts. The two gain blocks containing the matrix A_ESO and the vector [K 1] must be configured as matrix multiplications. If no control signal limitation is desired, the saturation block can be parameterized to ±Inf

Listing 10.1 MATLAB initialization code for the ADRC implementation in Fig. 10.1

```
% Tuning parameters
b0 = 1; omegaCL = 2*pi*0.2; kESO = 5;
% Discrete-time controller/observer gains, cf. Table A.5.1 on page 198
Ts = 0.1;
zCL = exp(-omegaCL*Ts); zESO = exp(-kESO*omegaCL*Ts);
K = [(1-zCL)^2/Ts^2,   (4-(1+zCL)^2)/(2*Ts)];
L = [(1-zESO^3);  3/(2*Ts)*(1-zESO)^2*(1+zESO);  1/Ts^2*(1-zESO)^3];
% Compute ESO matrix/vector A and b for N = 2, cf. Table A.5.2 on page 198
A_ESO = [1-L(1),   Ts-L(1)*Ts,    1/2*Ts^2-1/2*L(1)*Ts^2; ...
         -L(2),    1-L(2)*Ts,     Ts-1/2*L(2)*Ts^2; ...
         -L(3),    -L(3)*Ts,      1-1/2*L(3)*Ts^2];
B_ESO = [1/2*b0*Ts^2 - 1/2*L(1)*b0*Ts^2; ...
         b0*Ts-1/2*L(2)*b0*Ts^2; ...
         -1/2*L(3)*b0*Ts^2];
```

As can be seen from Fig. 10.1, the resulting ADRC block is a direct replication of Fig. A.5.1. The saturation block must be parameterized with the allowed output range of the control signal (u_{min} and u_{max}). The scalar, vector, and matrix gain blocks are making use of variables b0 (scalar, critical gain parameter b_0 to be obtained from plant modeling), K (row vector containing the controller gains k^T), L (column vector of observer gains l), A_ESO (system matrix A_{ESO} of the observer), and B_ESO (column vector, input gain b_{ESO} of the observer).

The initialization script required to set up the necessary matrices and vectors from ADRC's tuning parameters is given in Listing 10.1. User-defined values are the critical gain parameter b_0, the two bandwidth tuning parameters ω_{CL} and k_{ESO}, and the sampling interval T_s. Controller and observer gains k^T and l as well as the observer matrix/vector A_{ESO} and b_{ESO} are then automatically being computed exactly as given in Tables A.5.1 and A.5.2 for the order $N = 2$.

10.2 Implementation in C Programming Language

A second important option for implementing discrete-time ADRC is to manually write source code. Here, we will make use of the C programming language, the *lingua franca* of embedded systems, which allows to target practically every available processor or microcontroller platform.

10.2.1 State-Space Form

As a first example, we want to demonstrate how the discrete-time state-space form of ADRC, summarized in Appendix A.5, can be implemented in C source code, and also pay attention to practical aspects discussed in Chap. 9. A possible implementation of ADRC structure as shown in Fig. A.5.1 for $N = 1$ comprises the following steps.

Data Structure: We begin with the declaration of a data structure holding controller parameters and internal state variables. Inspection of (A.5.1) and (A.5.2) for $N = 1$ reveals that we need two state variables (for $\hat{x}_1$ and $\hat{x}_2$), and ten coefficients (controller and observer gains, b_0, observer matrix/vector elements). We will do that with one single `struct` as shown in Listing 10.2, also including parameters for limiting the controller output. A user will then have to instantiate this `struct`, and pass the instance to further C functions implemented below. A more refined approach separating coefficients and state variables would, of course, also be possible.

Listing 10.2 Discrete-time state-space implementation of first-order ADRC: definition of a combined data structure to hold controller parameters and states

```
struct ADRC1_SS
{
    /* coefficients of the state-space matrix A and vector b */
    double AESO11, AESO12, AESO21, AESO22, bESO1, bESO2;
    /* controller and observer gains, critical gain parameter */
    double k1, l1, l2, b0;
    /* parameters of the control signal limitation */
    double u_min, u_max;
    /* observer state variables */
    double x1, x2;
};
```

Parameter Tuning: Controller and observer gains can be obtained using *bandwidth parameterization*, as given in Table A.5.1. Based on that, the observer matrix/vector elements can be computed as listed in Table A.5.2. We therefore implement a C function that computes the coefficients based on tuning values ω_{CL}, k_{ESO}, as well as b_0 and T_s. Comparing the resulting Listing 10.3 with these tables confirms that the given equations can indeed be transformed into C source in a rather straightforward manner.

Listing 10.3 Discrete-time state-space implementation of first-order ADRC: initialize coefficient values using bandwidth parameterization

```
void ADRC1_SS_init_coefficients(struct ADRC1_SS * const adrc,
    double Tsample, double b0, double omegaCL, double kESO)
{
    /* locations in z-domain for pole placement */
    double zCL = exp(-omegaCL*Tsample);
    double zESO = exp(-kESO*omegaCL*Tsample);
    /* controller and observer gains, cf. Table A.5.1 (page 198) */
    adrc->k1 = (1.0 - zCL) / (Tsample);
    adrc->b0 = b0;
    adrc->l1 = 1.0 - zESO*zESO;
    adrc->l2 = (1.0 - zESO)*(1.0 - zESO) / Tsample;
    /* observer matrix AESO and vector bESO, cf. Table A.5.2 (page 198) */
    adrc->AESO11 = 1.0 - adrc->l1;
    adrc->AESO12 = Tsample*(1.0 - adrc->l1);
    adrc->AESO21 = -adrc->l2;
    adrc->AESO22 = 1.0 - Tsample*adrc->l2;
    adrc->bESO1 = b0*Tsample*(1.0 - adrc->l1);
    adrc->bESO2 = -b0*Tsample*adrc->l2;
}
```

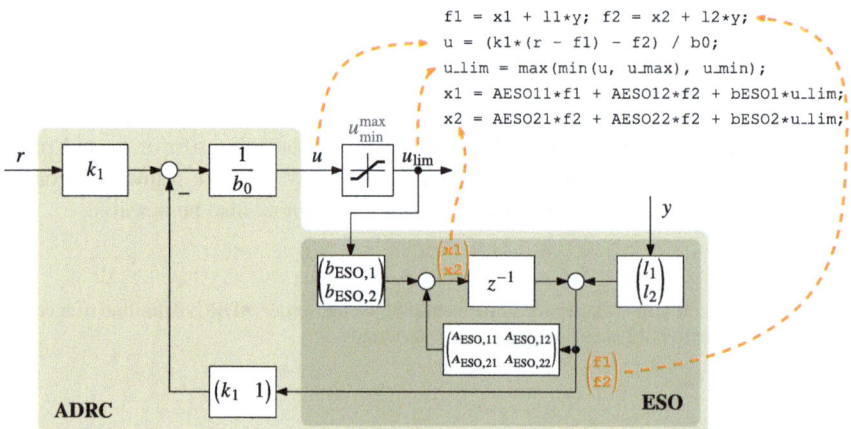

Fig. 10.2 Block diagram of the state-space form of first-order discrete-time ADRC, including an output magnitude limitation block, with C source code lines of a possible implementation of output and internal update equations of the controller

Control Law and Observer Update: To understand the implementation of the control law, we have redrawn Fig. A.5.1 for $N = 1$ in Fig. 10.2, introducing required state variables (x1 and x2). To abbreviate the resulting code, some intermediate variables (named f1 and f2 here) were introduced. These are the values fed back through the state-feedback control law, and also needed to update the state variables, i.e. the storage variables of the unit delay. Figure 10.2 links all required source code lines with their origin in the block diagram. As discussed in Sect. 9.1, every practical controller should limit its output and be equipped with anti-windup measures. We therefore added an output magnitude limitation to the controller, which was already

10.2 Implementation in C Programming Language

apparent from the block diagram. The resulting C function for computing a new, limited controller output value can be found in Listing 10.4.

Listing 10.4 Discrete-time state-space implementation of first-order ADRC: compute new limited controller output value and update internal states

```
double ADRC1_SS_compute_output(struct ADRC1_SS * const adrc,
       double r, double y)
{
    /* enhance internal state variables with current output measurement */
    double f1 = adrc->x1 + adrc->l1*y;
    double f2 = adrc->x2 + adrc->l2*y;
    /* compute controller output and apply magnitude limitation */
    double u = (adrc->k1*(r - f1) - f2) / adrc->b0;
    double u_lim = (u > adrc->u_max) ? adrc->u_max
                    : ((u < adrc->u_min) ? adrc->u_min : u);
    /* prepare internal state variables for next time step */
    adrc->x1 = adrc->AESO11*f1 + adrc->AESO12*f2 + adrc->bESO1*u_lim;
    adrc->x2 = adrc->AESO21*f1 + adrc->AESO22*f2 + adrc->bESO2*u_lim;
    return u_lim;
}
```

Customization: To add the ability to enable the controller from non-zero initial conditions for y and/or u, we implement the direct initialization strategy covered in Sect. 9.2 to ensure bumpless transfer. The resulting function is shown in Listing 10.5.

Listing 10.5 Discrete-time state-space implementation of first-order ADRC: initialize internal states for bumpless transfer from manual mode in a settled state $r = y$, cf. Sect. 9.2

```
void ADRC1_SS_init_states(struct ADRC1_SS * const adrc,
     double y, double u_initial)
{
    /* state x1: use measured plant output, minus the l1*y term */
    adrc->x1 = y - adrc->l1*y;
    /* state x2: total disturbance for bumpless transfer, minus l2*y */
    adrc->x2 = -adrc->b0*u_initial - adrc->l2*y;
}
```

This concludes our example implementation of first-order discrete-time ADRC in its state-space form. A user of this implementation will have to:

- instantiate the data structure from Listing 10.2;
- initialize the coefficients ("tune the controller") by calling the function in Listing 10.3;
- initialize the internal state variables before being able to bumplessly enable the controller using the function from Listing 10.5;
- and, periodically (once every T_s), call the update function in Listing 10.4 to compute the next controller output from the most recently measured plant output, before feeding the controller output to the actuator of the user's application.

10.2.2 Dual-Feedback Transfer Function Form

In a second example, we want to demonstrate how a transfer function variant of ADRC can be transferred from its block diagram to C source code. We pick the dual-feedback transfer function form in its minimum-footprint realization given in Fig. A.7.2 for $N = 1$, and implement it with the following steps.

Data Structure: By inspection of Fig. A.7.2 it becomes apparent that two state variables (for the two unit delay blocks) and seven coefficients are needed. Similar to the previous example, we will declare one single struct for parameters and internal state variables. The result is shown in Listing 10.6, also including parameters for the controller output limitation.

Listing 10.6 Discrete-time dual-feedback transfer function implementation of first-order ADRC: definition of a combined data structure to hold controller parameters and states

```
struct ADRC1_DFTF
{
    /* coefficients of the discrete-time transfer functions */
    double alpha1, alpha2, beta0, beta1, gamma0, gamma1, k1_b0;
    /* parameters of the control signal limitation */
    double u_min, u_max;
    /* state variables */
    double x1, x2;
};
```

Parameter Tuning: The transfer function coefficient values can be obtained using *bandwidth parameterization*, as given in Table A.7.2. The C function in Listing 10.7 computes these coefficients based on tuning values ω_{CL}, k_{ESO}, as well as b_0 and T_s. Once more, a comparison with Table A.7.2 reveals that the given equations can easily be transformed into C source code.

Listing 10.7 Discrete-time dual-feedback transfer function implementation of first-order ADRC: initialize coefficient values using bandwidth parameterization

```
void ADRC1_DFTB_init_coefficients(struct ADRC1_DFTF * const adrc,
        double Tsample, double b0, double omegaCL, double kESO)
{
    double zCL = exp(-omegaCL*Tsample);
    double zESO = exp(-kESO*omegaCL*Tsample);
    /* coefficient values for bandwidth param., cf. Table A.7.2 (page 203) */
    adrc->alpha1 = -2.0*zESO;
    adrc->alpha2 = zESO*zESO;
    adrc->beta0  = (zCL*zESO*zESO-2.0*zESO-zCL+2.0)/(b0*Tsample);
    adrc->beta1  = (2.0*zCL*zESO-2.0*zCL*zESO*zESO+zESO*zESO-1.0)/(b0*Tsample);
    adrc->gamma0 = zCL*zESO*zESO-2.0*zESO+1.0;
    adrc->gamma1 = zESO*zESO*(1.0-zCL);
    adrc->k1_b0  = (1 - zCL)/(b0*Tsample);
}
```

10.2 Implementation in C Programming Language

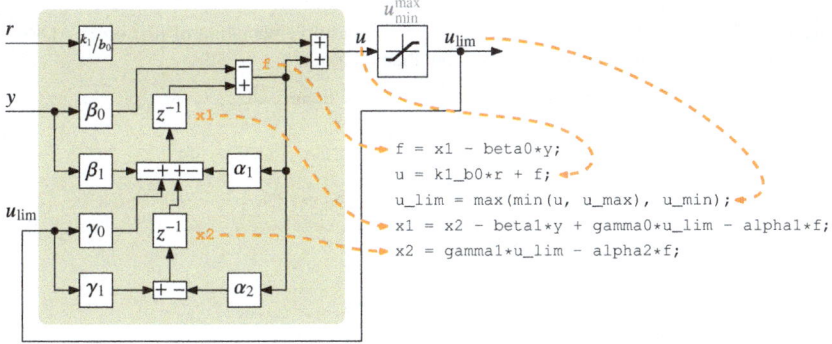

Fig. 10.3 Block diagram of the minimum-footprint dual-feedback transfer function form of first-order discrete-time ADRC (from Fig. A.7.2, now including an output magnitude limitation block), with C source code lines required to implement the output and internal update equations of the controller

Control Law: To implement the actual control law, one has to go over Fig. A.7.2 from top to bottom. This includes updating the two state variables (of the unit delay blocks), which we will name `x1` and `x2`. It is also useful to introduce an intermediate variable—named `f` here—which is being fed back through the α coefficients. To better support understanding the process of translating a block diagram to a programming language, Fig. 10.3 shows the block diagram annotated with C source code lines. A magnitude limitation was added to the controller output. The resulting C function for computing the controller output value is given in Listing 10.8.

Listing 10.8 Discrete-time dual-feedback transfer function implementation of first-order ADRC: compute new limited controller output value and update internal states, as shown in Fig. 10.3

```
double ADRC1_DFTF_compute_output(struct ADRC1_DFTF * const adrc,
        double r, double y)
{
    /* compute (unlimited) controller output */
    double f = adrc->x1 - adrc->beta0*y;
    double u = adrc->k1_b0*r + f;
    /* apply magnitude limitation to controller output */
    double u_lim = (u > adrc->u_max) ? adrc->u_max
                 : ((u < adrc->u_min) ? adrc->u_min : u);
    /* update internal states (unit delays) */
    adrc->x1 = adrc->x2 - adrc->alpha1*f - adrc->beta1*y + adrc->gamma0*u_lim;
    adrc->x2 = adrc->gamma1*u_lim - adrc->alpha2*f;
    return u_lim;
}
```

Customization: To bumplessly enable the controller from non-zero initial conditions for y and/or u, the direct initialization strategy from Sect. 9.2 is applied here, as well, resulting in Listing 10.9.

Listing 10.9 Discrete-time dual-feedback transfer function implementation of first-order ADRC: initialize internal states for bumpless transfer from manual mode

```
void ADRC1_DFTF_init_states(struct ADRC1_DFTF * const adrc,
    double y, double u_initial)
{
    /* initialize states for bumpless transfer, cf. Sect.9.2 */
    double f = u_initial - adrc->k1_b0*y;
    adrc->x2 = adrc->gamma1*u_initial - adrc->alpha2*f;
    adrc->x1 = adrc->x2 - adrc->alpha1*f - adrc->beta1*y
              + adrc->gamma0*u_initial;
}
```

This concludes our implementation. From a user's perspective, this implementation is to be treated in exactly the same manner as the state-space form covered in Sect. 10.2.1. Using these functions requires:

- instantiating the data structure from Listing 10.6;
- initializing the coefficients (= tuning the controller) by calling the function in Listing 10.7;
- initializing the internal state variables before being able to bumplessly enable the controller using the function from Listing 10.9;
- and, periodically (once per sampling interval T_s), calling the update function in Listing 10.8 to compute the next controller output from the most recently measured plant output. The resulting controller output must then be fed to the actuator of the user's application.

Open Access This chapter is licensed under the terms of the Creative Commons Attribution 4.0 International License (http://creativecommons.org/licenses/by/4.0/), which permits use, sharing, adaptation, distribution and reproduction in any medium or format, as long as you give appropriate credit to the original author(s) and the source, provide a link to the Creative Commons license and indicate if changes were made.

The images or other third party material in this chapter are included in the chapter's Creative Commons license, unless indicated otherwise in a credit line to the material. If material is not included in the chapter's Creative Commons license and your intended use is not permitted by statutory regulation or exceeds the permitted use, you will need to obtain permission directly from the copyright holder.

Chapter 11
Application Examples

Abstract Having already introduced various extensions and modifications of the ADRC structure and its discrete-time variants for software-based implementation, we are ready to put it to work! To show how ADRC performs in actual applications, here we consider two laboratory testbeds representative of a spectrum of real control problems one can encounter in engineering practice. For those application examples, custom-made ADRC-based solutions are devised, designed, and deployed. We make an arbitrary choice to use the previously derived discrete-time state-space ADRC in the first example and its transfer function form for the implementation in the second example. For each of the considered cases, we will show similar steps including a brief plant description, formulation of the control problem, implementation and commissioning of the selected ADRC (based on already introduced "cooking recipes"), initial validation in simulation, and finally physical validation on target hardware. The examples will differ vastly regarding not only their dynamics but also the method chosen for plant modeling, covering both an experimentally driven approach and a case based on theoretical and numerical analyses.

11.1 Heater Temperature Control

A temperature controller is often used in scenarios where an object is required to be heated, cooled, or both, to remain at the target temperature, regardless of the changing environment around it. Many industrial processes require a precisely kept temperature to ensure product quality, prevent equipment damage, improve efficiency, and maintain a safe working environment. In this example, we utilize Temperature Control Lab (TCLab), which is a low-cost and commercially available Arduino-based education module.[1] It emulates a real, multi-input multi-output process control plant, where two transistors (working as heaters) and two thermistor temperature sensors

---

[1] The TCLab board is available from https://apmonitor.com/pdc/index.php/main/PurchaseLabKit where one can also find more technical details on it.

serve as the plant's inputs and outputs, respectively. The usually considered control objective in TCLab is to apply power to the heaters to keep desired temperatures. The two heater units are deliberately placed by the designers on the printed circuit board in close proximity to each other to allow heat transfer by convection and thermal radiation. This makes the control of TCLab additionally challenging as the operation of one heater influences the other, and vice versa, thus creating a cross-coupling effect. At the same time, some of the heat is also transferred away from the device to the surroundings. In our case, we will be considering TCLab as an SISO control problem, where the control objective is to govern the temperature of the first heater, and the influence of the second heater is treated as an unmodeled, external disturbance to the first one.

The input signals to the TCLab are normalized values of $0\% \ldots 100\%$, corresponding to $0\,\text{W} \ldots 10\,\text{W}$, which is the PWM-controlled power range of the heaters. The output signals are expressed, for user's convenience, in degree Celsius but originate from analog voltages of the temperature sensors captured using analog-to-digital converters. Their accuracy is $\pm 1\,°\text{C}$ at best. A USB cable connects the Arduino to a computer allowing serial data communication. An overview of the TCLab device setup can be seen in Fig. 11.1.

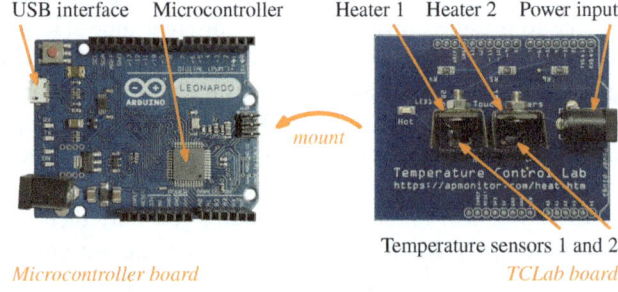

Fig. 11.1 Target hardware for the temperature control example: TCLab board, available from APMonitor, to be attached to an Arduino development board

Plant Modeling

Although the ADRC structure has been shown in Chap. 5 to have a considerable level of robustness against parametric and structural uncertainties and external disturbances, the procedure of system mathematical modeling should not be deemed unnecessary and upfront discarded. The benefit of even a rough plant modeling for the purpose of ADRC design was shown before in Sect. 6.1. Therefore, for the TCLab, we will follow the straightforward modeling procedure from Sect. 5.1 of experimentally capturing the system's core behavior using its step response. To avoid confusion, Table 11.1 contains the established terminology and notation.

11.1 Heater Temperature Control

Table 11.1 Notation used in the TCLab example with real components seen also in Fig. 11.1

Element	Assigned role	Symbol
Heater 1	Control input	u
Temperature sensor 1	Controlled variable	y
Temperature sensor 1 (model)	Controlled variable	y_m
Heater 2	Disturbance input	u_d
Temperature sensor 2	Disturbance output	y_d

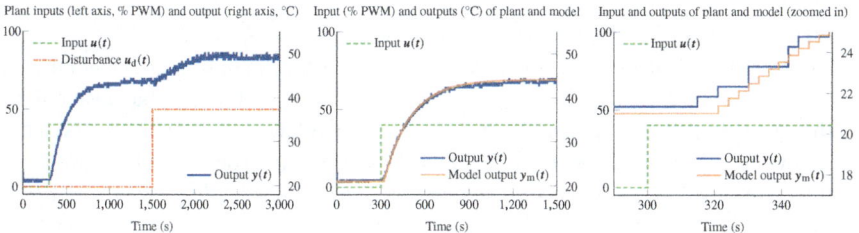

Fig. 11.2 Open-loop step responses of TCLab (left figure, showing the effects of both control and disturbance inputs), leading to the model (11.1). Plant and model are compared in the center figure. Note that the model output is shifted by the ambient temperature for comparability. To make the dead time more visible, a zoomed-in detail of the model comparison is repeated in the right figure

To get the step response of TCLab, we use a stepwise change in the input signal $u(t) = 40\% \cdot \sigma(t)$ at time $t = 300$ s. The sampling interval is set to $T_s = 3$ s. The results of such an open-loop test are presented in Fig. 11.2. One can notice that the process exhibits mostly first-order low-pass behavior, similar to the one discussed in Sect. 5.1, with an approximate dominant time constant $T \approx 182$ s and gain $K \approx 0.58 \frac{°C}{\%\text{Heater}}$. In this case, the process also has a dead time of $T_d \approx 18$ s. Therefore, based on the collected data, the rough plant model of the TCLab is chosen as

$$T \cdot \frac{dy_m(t)}{dt} = -y_m(t) + K \cdot u(t - T_d)$$

or alternatively in a transfer function form, similar to (9.5):

$$P(s) = \frac{K}{Ts + 1} \cdot e^{-T_d s} = \frac{0.58 \frac{°C}{\%\text{Heater}}}{182 \text{ s} \cdot s + 1} \cdot e^{-18 \text{ s} \cdot s}. \quad (11.1)$$

To assess the level of model fitting, the step response of (11.1), denoted as $y_m(t)$, to the same stepwise input signal $u(t)$ is added to Fig. 11.2. It should be that (11.1) is a rough description of the actual system, not including phenomena like sensor noise or cross-coupling from the other heater, which we use as a disturbance input. We could follow the path of more accurate modeling but that would require investment of additional resources, like time, measurement capabilities, etc. Instead, we will deliberately use the simple SISO model (11.1) to show that ADRC can effectively work even if only limited information about the controlled system is available.

ADRC Design

To develop an ADRC scheme for the considered TCLab system, let us make use of the already possessed knowledge and follow the "cooking recipe" for the discrete-time state-space form of linear ADRC given on page 130:

1. *Plant modeling:* The plant order of $P(s)$ in (11.1) is $N = 1$, calling for a first-order ADRC. From the step response, we already obtained plant parameter estimations, namely $K \approx 0.58\,\frac{°C}{\%_{\text{Heater}}}$, $T \approx 182$ s, and $T_d \approx 18$ s. Recalling Table 3.1, we therefore determine the *critical gain parameter* as $b_0 = \frac{K}{T} \approx 0.003\,\frac{°C}{\%_{\text{Heater}} \cdot s}$.

2. *Controller and observer tuning:* We are not given certain design goals and therefore use a settling time $T_{\text{settle},98\%} \approx 300$ s to be at least somewhat faster than the open-loop step response. With (3.21) we calculate $\omega_{\text{CL}} \approx 0.013$ rad/s. This allows for a practically acceptable signal-to-noise ratio. The controller gain k_1 is then computed using the general design equation for *bandwidth parameterization* given by (8.13), or through referring to Table A.5.1, which results in $k_1 \approx 0.0128$. Given the noise levels in the open-loop measurements collected so far, we select the relative observer bandwidth factor to a conservative value of $k_{\text{ESO}} = 3$. To compute the observer gains, we put ω_{CL} and k_{ESO} into the general design Eq. (8.6), or once more refer to Table A.5.1, obtaining $l_1 \approx 0.2094$ and $l_2 \approx 0.0041$.

3. *Implementation:* In our model-based implementation, we use the discrete-time state-space form of ADRC for plants with dead time, depicted in Fig. 9.9. It requires the discrete-time control law (8.1) and the discrete-time ESO equation (9.6) to be implemented. Since we roughly know the TCLab dead time to be $T_d \approx 18$ s, the observer synchronization method from Sect. 9.4 is applied with the control signal being delayed by $n_d = 6 \approx \frac{T_d}{T_s}$ samples. The sampling interval used for the discrete implementation is $T_s = 3$ s.

4. *Customization:* Our practical controller is extended by a discrete-time magnitude and rate limitation add-on block from Sect. 9.1. Output limits $u_{\min} = 0\,\%$, $u_{\max} = 100\,\%$ follow from TCLab's normalized heater power range. The rate limitation is, primarily for demonstration purposes, configured with values $\Delta u_{\min/\max} = \pm 1\,\frac{\%}{s}$. Finally, since the TCLab starts its operation at room temperature (most likely not zero), we apply the *direct initialization* from Sect. 9.2 and use an initial ambient temperature measurement as a starting condition to bumplessly switch on both observer and controller.

Given the above design, the ADRC scheme for the TCLab can be represented with a block diagram seen in Fig. 11.3.

Validation in Simulation

Before we deploy the above ADRC on the actual TCLab, we will first test it in simulation, which is overall a good practice in control engineering that helps with concept validation and code debugging. Here, the control objective is simply to keep the reference temperature of the first heater by manipulating its input power. We can

11.1 Heater Temperature Control

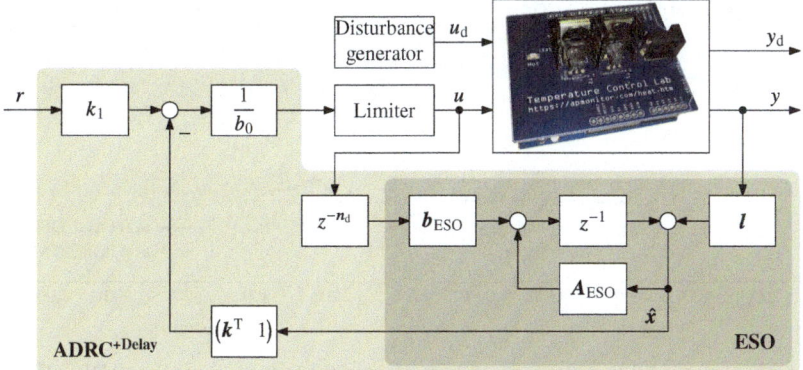

Fig. 11.3 Discrete-time ADRC for TCLab, state-space implementation, enhanced by a control signal limitation block and delayed control signal to synchronize the observer with the time-delayed plant. Note that we, in this example, denote the limited controller output simply as u

only test here an SISO control system since the influence of the second heater was assumed to be a disturbance; hence we do not have a model for it. The presence of dead time in the TCLab system is an excellent opportunity to apply the ESO dead time synchronization method described in Sect. 9.4. Therefore, we test two ADRC designs, one with synchronization turned on (denoted as "sync. ON") and the other without synchronization ("sync. OFF"). The test consists of a stepwise change of the reference signal $r(t)$ to 40 °C at time $t = 0$ s, starting from a simulated ambient temperature of 23 °C. The results in Fig. 11.4 show the closed-loop behavior, with $y(t)$ approaching $r(t)$. It is nicely evident that if information about the dead time is available, it can be put to good use, leading to reduced efforts and oscillations in the controller output signal.

Validation in Experiment

Since our initial control concept for the TCLab temperature control was successfully tested in a simulated environment, we can now feel more comfortable deploying it on the real device. The discretized ADRC algorithm is implemented on the Arduino board of the TCLab. The software implementation of the "cooking recipe" is a straightforward procedure as it is shown in Sect. 10.2.1, extended by the delay synchronization. This time, the control objective is to keep the reference temperature of the first heater by manipulating its input power while remaining insensitive against the interferences from the second heater, which acts as an unmodeled, external disturbance. Similar to the simulation validation, we are interested here in how the dead time synchronization method from Sect. 9.4 works on the control performance; hence we test again the two cases "sync. ON" and "sync. OFF." As for the details of the test itself, at time $t = 0$ s, a stepwise change of the reference signal $r(t)$ to 40 °C occurs followed by a stepwise change at $t = 1200$ s of the disturbance input signal

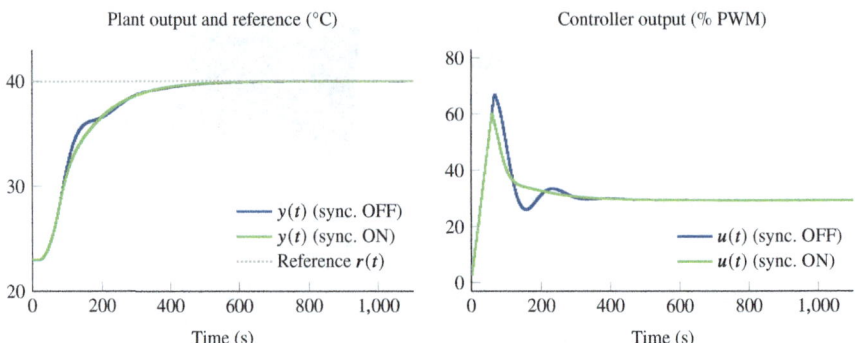

Fig. 11.4 Simulation validation of the closed-loop step response using discrete-time ADRC. Even though similar regulation performance can be seen for the controlled variable y of the two tested designs, ADRC without dead time synchronization in the ESO ("sync. OFF") generates a visibly larger and more oscillatory control signal u than the scenario with synchronization turned on ("sync. ON"), which is not desirable, e.g., from an energy usage point of view

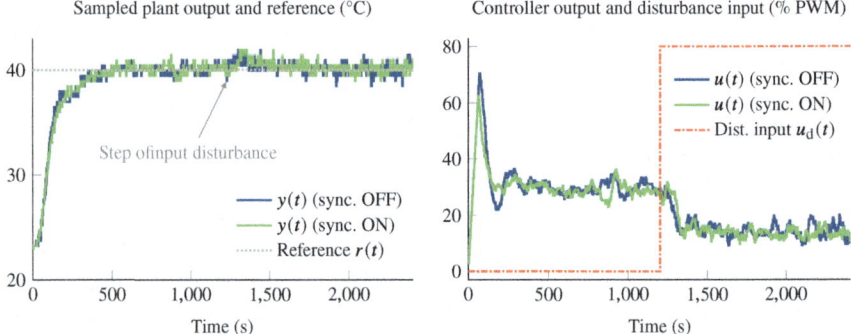

Fig. 11.5 Experimental validation of the closed-loop step response and disturbance rejection using discrete-time ADRC (without and with dead time synchronization). A user-defined disturbance from the second heater $u_d(t) = 80\,\%$ becomes effective starting from $t = 1200\,\text{s}$. In contrast to the open-loop case in Fig. 11.2, the effect of the disturbance is now being compensated very well by the controller, as visible in the controller outputs

$u_d(t)$ to $80\,\%$. From the results in Fig. 11.5, one can see that the control objective is being realized in a practically acceptable manner with the output $y(t)$ exhibiting robustness to the influence of the second heater, which is especially visible in the disturbance countering measures by the control signal $u(t)$.

11.2 DC-DC Converter Voltage Control

Let us now dive into an application domain with significantly smaller time constants: power electronics. More specifically, the problem of output voltage control in a step-down (buck) DC-DC converter shall be addressed. Once more, we consciously decided to target an affordable, scaled-down experimental platform. This shall encourage readers to replicate these results and lower the barriers to getting started with own experiments of ADRC in a real-world control loop. Our experimental setup consists of a development board with a 32-bit microcontroller unit (MCU) attached to a second board with the DC-DC converter and a switchable load resistor, as shown in Fig. 11.6. Both are available commercially.[2]

This second application example once more aims to implement an ADRC control loop using the tools introduced in this book. The tuning parameters achieved may then serve as a solid starting point to fine-tune the loop with respect to certain control-related optimization criteria, but this is not our goal. Instead, we intend to emphasize the fact that a working implementation of ADRC can be found without trial and error, by following the guidelines presented in previous chapters.

Fig. 11.6 Target hardware for the voltage control example: DC-DC converter board attached to a microcontroller development board, both available from Texas Instruments

Plant Modeling

The converter shall be operated in the so-called voltage mode, i.e., a single control loop regulating the output voltage. Control signal will be the duty cycle, i.e., the relative on-time of the converter's high-side switch. For the intended ADRC design, we therefore must determine a model with input *duty cycle* and output *output voltage*.

---

[2] The microcontroller development board (F28069M LaunchPad) employed in this is available from https://www.ti.com/tool/LAUNCHXL-F28069M, the buck converter add-on board (BoosterPack) from https://www.ti.com/tool/BOOSTXL-BUCKCONV. *LaunchPad* and *BoosterPack* are trademarks of *Texas Instruments*.

Fig. 11.7 Simplified schematic of the synchronous buck DC-DC converter. Inductor ESR and current sense resistor have been combined ($r_L + r_{CS}$)

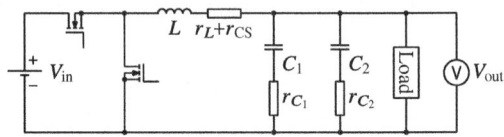

A simplified schematic of the step-down (buck) DC-DC converter is shown in Fig. 11.7. Compared to a "textbook" converter, mixed-type output capacitors are being used here: higher capacitance electrolytic capacitors with large equivalent series resistance (ESR) combined with low-capacitance low-ESR ceramic capacitors. As pointed out in dedicated literature, this can considerably complicate the process of modeling and control of the plant.[3]

In frequency domain, a model for the output voltage control task can be derived from the voltage divider formed by the load and the power stage components:

$$P(j\omega) = \frac{y(j\omega)}{u(j\omega)} = \frac{V_{out}(j\omega)}{\text{Duty}(j\omega)} = V_{in} \cdot \frac{Z(j\omega)}{Z(j\omega) + (r_L + r_{CS} + j\omega L)} \quad (11.2)$$

$$\text{with} \quad Z(j\omega) = R_{\text{Load}} \parallel \left(r_{C1} + \frac{1}{j\omega C_1}\right) \parallel \left(r_{C2} + \frac{1}{j\omega C_2}\right).$$

Table 11.2 Component values used for creating the plant model of the DC-DC converter

Description and symbol	Value	Description and symbol	Value
Capacitance C_1	330 μF	ESR r_{C1}	150 mΩ
Capacitance C_2	66 μF	ESR r_{C2}	2 mΩ
Inductance L	4.8 μH	ESR $r_L + r_{CS}$ (current sense)	40 mΩ
Load resistance R_{Load} (base load)	7.5 Ω	Load resistance R_{Load} (full load)	1.6 Ω

Analytically, this would result in a relatively convoluted transfer function. Remembering that, apart from the plant order N, ADRC requires only one plant model parameter b_0, we do not need to be afraid, however. With the component values listed in Table 11.2, Bode diagrams of the plant dynamics can be computed numerically for the base and full load cases. As shown in Fig. 11.8, one can extract the required information easily from these diagrams. In a frequency interval relevant to our control design (aiming for bandwidth in the kilohertz range), dynamics are clearly of order $N = 2$. Using the frequency-domain method introduced in Sect. 5.1.2, an approximate value of b_0 can be obtained with $b_0 \approx 150000^2 \frac{V}{s^2}$. This already concludes our efforts regarding plant modeling of this converter.

---

[3] We would like to refer the reader to https://www.ti.com/lit/pdf/slvae26 and https://www.powerelectronictips.com/how-mixed-type-output-capacitors-affect-dcdc-converter-stability for more details on this subject.

11.2 DC-DC Converter Voltage Control

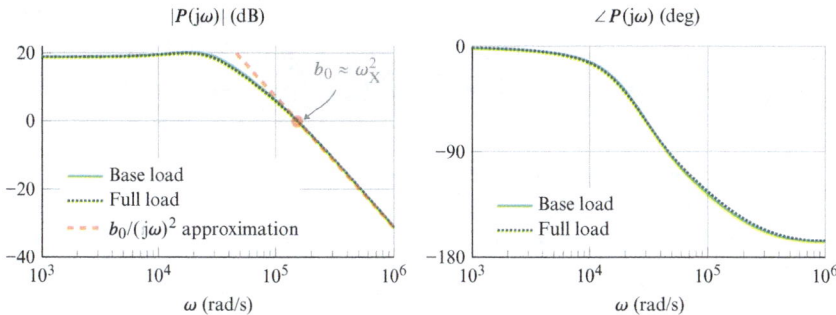

Fig. 11.8 Bode diagrams of the plant dynamics for the two possible cases (base load and full load). As expected, the differences are negligible. In the frequency range relevant to the control of this plant, the dynamics predominantly exhibit second-order behavior. Using the approach introduced in Sect. 5.1.2, we can find the plant models' critical gain parameter b_0 from the crossover frequency of a straight-line approximation of the magnitude plot

ADRC Design

The switching frequency of the DC-DC converter is 200 kHz. Updating the control loop every switching period will allow us to aim for a large closed-loop bandwidth but require a computationally efficient implementation. We will therefore use the low-footprint dual-feedback transfer function form of ADRC described in Sect. 8.3 and follow the "cooking recipe" from page 138.

1. *Plant modeling:* As depicted in Fig. 11.8, the plant is of order $N = 2$, with an approximate critical gain parameter value of $b_0 = 150000^2 = 22.5 \cdot 10^9$.
2. *Controller and observer tuning:* Aiming for a 98 % settling time of $T_{\text{settle}} = 200\,\mu\text{s}$ leads, when put into (3.21), to a desired closed-loop bandwidth $\omega_{\text{CL}} = 29000\,\text{rad/s}$. To minimize the effects of measurement noise, the observer bandwidth will be chosen conservatively with $k_{\text{ESO}} = 3$. Furthermore, we will make use of half-gain tuning for the observer as described in Sect. 9.3.
3. *Implementation:* As already mentioned, the minimum-footprint transfer function variant shown in Fig. 8.9 will be used, with a sampling interval $T_s = \frac{1}{200\,\text{kHz}} = 5\,\mu\text{s}$.
4. *Customization:* A magnitude limitation of the controller output to the possible interval of the PWM duty cycle [0 %, 100 %] is added, i.e., $u_{\min} = 0$ and $u_{\max} = 1$.

Figure 11.9 gives an overview of the control loop in block diagram form. The controller will be implemented on the microcontroller shown in Fig. 11.6.

Validation in Simulation

A simulation model of the converter as depicted in Fig. 11.7 using the parameter values from Table 11.2 was set up to analyze the closed-loop behavior for the output voltage control task. A small simulation scenario was designed, consisting of the following elements, which are executed after the output voltage is settled at $r = 2\,\text{V}$:

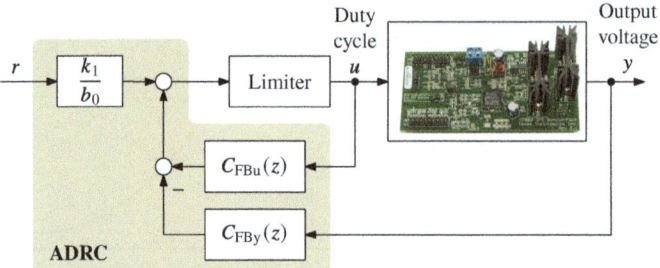

Fig. 11.9 Discrete-time ADRC for the DC-DC converter. The dual-feedback transfer function implementation is being used, enhanced by a control signal limitation block. In this example, we also denote the limited controller output simply as u

- A step of the reference signal r to a new desired output voltage value 2.2 V occurring at $t = 2$ ms
- A load step at $t = 4$ ms (activating full load on the converter board)
- A load dump at $t = 6$ ms (reduction to the base load resistor)

Figure 11.10 shows the simulation results. To examine the controller's sensitivity to measurement noise, a Gaussian noise signal was added to the output voltage with $\sigma = 2$ mV. While not exactly achieving a settling time of 200 μs in the transient after the reference step, these results obtained by only following the "cooking recipe" without further tweaking are already very well usable. While the performance could, as always, still be refined if necessary, we will therefore now directly transfer this design into practice.

Validation in Experiment

The controller implementation on the MCU target platform was carried out in the C programming language. For selected ADRC form, this was done as described in detail in Sect. 10.2.2, but for the second-order case and—owing to the properties of the MCU target platform—only using single-precision floating point numbers. The implementation made use of the high-resolution PWM capabilities of the selected MCU and, at a switching frequency of 200 kHz, used a PWM dead-band of approximately 60 μs.

Measurement results are presented in Fig. 11.11. They were recorded in a buffer on the MCU and later transferred to a host PC, thereby revealing the exact values as measured and computed by the MCU. The virtually identical reference step transient confirms that the design was successfully transferred to the target platform.

Summarizing the DC-DC converter example, we can state that a controller design which can serve as a very solid starting point for further optimization could directly be obtained by following the "cooking recipe" and using the tools provided here. Exactly that was our intention when writing this book: providing a direct path from principles to practice.

11.2 DC-DC Converter Voltage Control

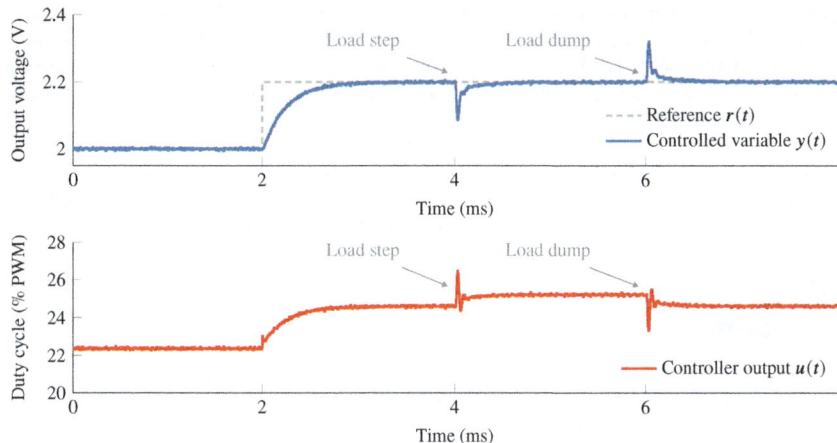

Fig. 11.10 Simulation results of DC-DC converter voltage control using second-order discrete-time ADRC. At time $t = 2$ ms, a reference signal step from 2 V to 2.2 V occurs. Between $t = 4$ ms and $t = 6$ ms, the converter operates in its full load condition

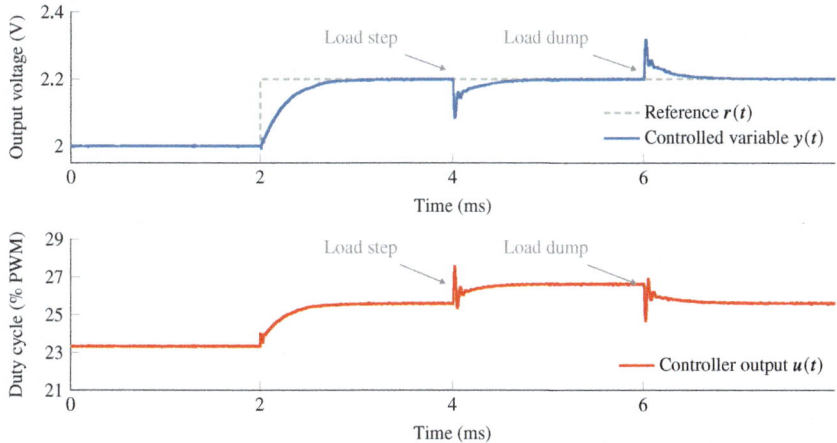

Fig. 11.11 Measurement results from the DC-DC converter target hardware, when running through the same test scenario (reference signal step, load step, and load dump). Overall, a very good replication of the simulation data was achieved. Settling time and required control efforts following the disturbance steps are (if only slightly) higher than in simulation. The reference signal step transient is almost indistinguishable from its counterpart in Fig. 11.10

Open Access This chapter is licensed under the terms of the Creative Commons Attribution 4.0 International License (http://creativecommons.org/licenses/by/4.0/), which permits use, sharing, adaptation, distribution and reproduction in any medium or format, as long as you give appropriate credit to the original author(s) and the source, provide a link to the Creative Commons license and indicate if changes were made.

The images or other third party material in this chapter are included in the chapter's Creative Commons license, unless indicated otherwise in a credit line to the material. If material is not included in the chapter's Creative Commons license and your intended use is not permitted by statutory regulation or exceeds the permitted use, you will need to obtain permission directly from the copyright holder.

Chapter 12
Postlude: A Look Ahead

Abstract Here we deal with the future, both nearest and that a bit further. First, supported by the body of knowledge we have put forward with this book, we make a case for using ADRC in the future. Then, we provide key takeaways on how to use ADRC moving forward and how this book can facilitate that. And even though we are heading toward the end of Part II, this is not the end of the book yet. We explain what substantial information is still coming and why it can be especially useful to returning readers. Here we also gather and list all the "cooking recipes" which are ready-to-use procedures to tune and implement all ADRC variants covered in the book. Finally, we take a look at ADRC in a broader sense and discuss possible avenues for its further development.

Why to Use ADRC in the Future?

Entering now the back end of the book, it is time to summarize its efforts. From the start, we wanted to encourage readers to try ADRC, and here we recap and reflect on the arguments that speak for its use in the future. Looking at the state of the art of ADRC and the hyper-growth of topics associated with it, we see that ADRC, as a technology, has reached its maturity phase. It has been in use for long enough that most of its initial faults and inherent problems have been removed or reduced by further development. It is relatively well established that we know what it can and cannot do, how it is related to classical control, and therefore we can at last realistically assess its strengths and limitations. Finally, it seems that the control community has accepted ADRC, and although it has not seen widespread practical use, its scientific background is rather clear. And this maturity of ADRC technology is one of the motivations behind our decision to write this book right now. So with everything we have shown so far in the book, what are the core arguments behind using ADRC in the future? First of all, it is not a magical, one-size-fits-all solution. It is, however, a solid practical linear control. Even though we now know that the linear ADRC is not much different than a PID and that you could tune a modified PID loop in the same way you tune ADRC (as we have shown in the book), it does come with built-in qualities that make it a good default choice when looking for a practical solution to tackle a real control problem. Based on the detailed analyses and visual comparisons we have conducted throughout the book, ADRC can be characterized by the following feature set:

- **Generally good out-of-the-box control behavior**: As we have shown, linear ADRC offers a relatively large level of robustness and adaptability while providing a practically acceptable ratio between necessary design effort and resultant performance. If one chooses an output-based variant of ADRC, then one gets a nicely critically damped tuning, which fits most solutions.
- **Easier tuning**: The use of *bandwidth parameterization* greatly simplifies the controller tuning process by minimizing the number of tuning gains and, by introducing the notion of bandwidth, also makes the tuning process more intuitive with its two practically appealing time- and frequency-domain interpretations. This makes ADRC more transparent than classical controllers. Compared to, for example, the time-tested PID, you still have some parameters to tune in the case of ADRC. But for ADRC it is conceptually a different task than finding PID parameters, which are more intertwined than the above parameters, and for ADRC we know more about which compromises one can expect when having certain parameter choices.
- **Easy to build in windup protection**: The structure of ADRC facilitates a simple, pragmatic, yet effective modification that results in gained windup protection, which is especially crucial in real systems. Although this feature is not magical and is rather a standard operation procedure in practice, it is convenient that in ADRC it is there and one does not have to think about it.
- **Low-footprint implementation**: The above core features do not cost extra computational resources. The implementation of linear ADRC is not much more demanding than that of a classical controller. If needed, there are specialized variants of ADRC (some of which we have covered in the book) that are explicitly dedicated to practical control systems that require high computational efficiency and low parameter footprint.
- **Customizability**: As we have shown in many instances throughout the book, ADRC is not a single set of equations. It can be straightforwardly tailored to a given application depending on many factors, like design requirements, characteristics of the controlled plant, or limitations of the actuators or the sensors. This flexibility in design is similar to what we see in, for example, PID with all of its iterations and customizations introduced over the years.

How to Use ADRC in the Future?

The use of ADRC always boils down to a certain amount of steps. From our experience, these mostly are the following four, which you could already recognize from the previously introduced "cooking recipes":

1. **Plant modeling**

    Plant relative order (N): In many instances, it is generally known or relatively straightforward to be somewhat accurately approximated by a low number, usually $N = 1$ or $N = 2$. Note: That is also the reason why our simulation and experiments in the book are first or second order as they are most common in practice.

    Critical gain parameter (b_0): There are different options for choosing this parameter. One is using a time-domain model, the concept we introduced in Sect. 5.1.1 and then applied to the first hardware example in Sect. 11.1. One can also use the frequency domain model, as we have shown in Sect. 5.1.2 and then use this concept for the second hardware example in Sect. 11.2. If you deal with a plant with simple first- or second-order low-pass behavior and have its transfer function model, then you could calculate b_0 as done in Table 3.1.

2. **Controller and observer tuning**

    Controller bandwidth (ω_{CL}): It is much easier in the case of ADRC methodology to have an idea from the application itself about its value (in other words how fast the settling time could be). This has been made very clear in the hardware examples we have provided where we dealt with two systems with significantly different time constants: one in hundreds of seconds and the other in milliseconds. Those values directly tell us about how big the searching space is for ω_{CL} and what is realistically possible in terms of system performance. Tuning has to also take into consideration factors limiting the bandwidth and related to the intrinsic characteristics of the system and the environment it operates in (e.g., input delay). Note: Such a system dynamics-related approach is much easier than trying to figure out, for example, PI parameters.

    Observer bandwidth (tuned indirectly via k_{ESO}): Recalling (3.22), the observer bandwidth results from selecting the scaling factor k_{ESO}. Interestingly, even if one keeps its value within a default range $k_{ESO} = 3, \ldots, 10$, there is a good chance you will obtain satisfactory performance. This results from an intuitive relation within ADRC that the inner, disturbance estimation loop (driven by the observer bandwidth) has to work faster than the feedback control loop (driven by the controller bandwidth). In practice, the bandwidth selection will always be affected by an upper bound resulting from the limitation of the controlled system (e.g., sensor noise), something we have shown in detail in Chap. 5. Note: For systems with N more-less known and for which the default value of k_{ESO} is expected to provide satisfying results, this could potentially mean a necessity of finding only b_0 and ω_{CL}.

3. **Implementation**

    In the book, we have presented several different ways (variants) of implementing ADRC, each having its own characteristics, but the above tuning applies to all variants we covered. To aid the readers with choosing a suitable variant, we provide an overview in Fig. 12.1, where three major questions need to be answered: continuous or discrete time? state-space or transfer function form? "classical" (output-based) or error-based ADRC? Note: Here we show just a selection of variants that, in our opinion, are most useful. In general, other variants can be used as well, and it is up to the user to decide what is the most fitting type of

ADRC implementation depending on the given scenario. The next step is the implementation of the actual code. Depending on the earlier selection of the most suitable variant, one can go with a continuous-time implementation or, what is more likely in practice, with a discrete-time one. In Chap. 10, we show in great detail how discrete-time variants can be deployed in software form using a model-based environment or with manual coding using C language.

4. **Customization**

    Finally, if there is a clear need to further improve the control design, the "standard" ADRC (as introduced in Chap. 3) could be tweaked. Due to the flexibility of ADRC in terms of its structure and parameters, it can take into consideration different practical aspects, for example, by using the extensions and modifications we have shown in the book. Augmenting ADRC with such add-ons can create tailor-made solutions to directly address certain important characteristics and limitations of the system, like plant dead time and measurement noise.

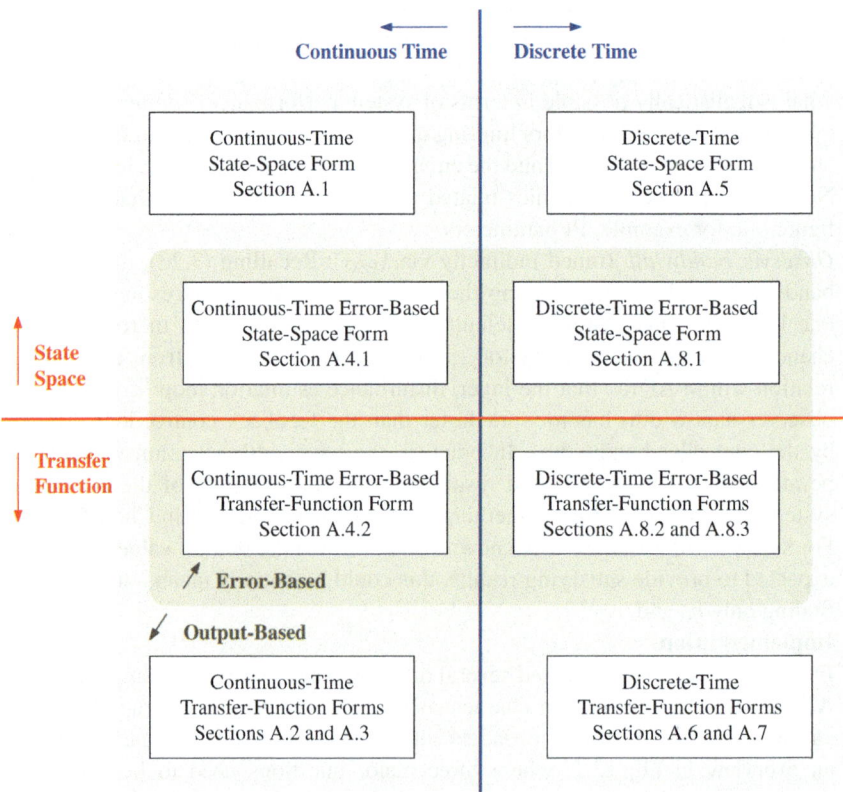

Fig. 12.1 Overview of ADRC variants discussed in this book and where to find their summary in the Appendix. To aid with the decision, three major questions to be answered have been visualized: continuous or discrete time? state-space or transfer function form? "classical" (output-based) or error-based ADRC?

How to Use This Book in the Future?

The graphical guide we started with all the way back in Fig. 1.1 can serve as a useful gateway for returning readers, allowing them to quickly jump back into whatever aspect is interesting to them or needs repeating:

- **Wanting to consolidate foundations**: If you want to solidify your understanding of the fundamental methodology of ADRC, its operating principle, basic tools, and methods, then you are invited back to the elemental Chaps. 2 and 3.
- **Wanting to deeper understand**: If you are looking to go beyond the fundamentals, then you can check Chaps. 4 and 5. There you will find how to design the most important variants of ADRC and how to parameterize and tune ADRC while understanding the practical role of tuning parameters on system performance. These parts also detail the derivation of ADRC core concepts and help to understand the characteristics of its components (in both time and frequency domains) and the connection between the ADRC methodology and classic controllers, which also helps to understand the strengths and limitations of ADRC, both theoretical and practical. We refer the readers seeking historical perspective and relevant bibliographical support for ADRC to Chap. 7.
- **Wanting to deploy in practice**: If you already understand ADRC and want to use it writing control software, then you can go to Chap. 10, where selected discrete-time ADRC variants (derived in Chap. 8) in software form are implemented using both a model-based environment and manually with C code. One can also go for a quick and easy Simulink-based implementation of ADRC with one of the software libraries shown in Appendix B.
- **Wanting a reference manual**: If you are an experienced control engineer and you know how to deploy things in practice, then just start applying ADRC using the prepared "cooking recipes" (Table 12.1) as we have done for examples in Chap. 11. For that, you will need "ingredients," and we provide those in a condensed way in the reference-style Appendix A. This cheat sheet could potentially be the only source material needed for your future ADRC implementations.
- **Wanting to tailor-make**: Knowing how to design and implement customized versions of ADRC, including some of its extensions and modifications and taking into account various practical aspects of the controlled systems may turn out to be very beneficial moving forward, especially when faced with nontrivial control problems and stringent control goals. Chapters 6 and 9 show a variety of ways that can aid designers with that.

The first four points happen to align exactly with the book's subtitle "from principles to practice," where, in this context, *principles* mean understanding and *practice* means deployment. With the last point, related to tailoring, we invite the reader to continue the journey by roaming freely and exploring ADRC beyond the rudimentary ideas and tools we have covered in the book.

Moving forward, the book can be easily revisited for consultation as a textbook, handbook, or reference, all to help different audiences achieve their desired goals.

Table 12.1 List of "cooking recipes" that are ready-to-use procedures to tune and implement linear ADRC. For each variant, the required set of equations is, along with a block diagram, provided in condensed form in Appendix A

Linear ADRC Variant	Location	Details
Continuous-Time State-Space Form	Page 42	Appendix A.1
Continuous-Time Transfer Function Form	Page 54	Appendix A.3
Continuous-Time Error-Based State-Space Form	Page 97	Appendix A.4.1
Continuous-Time Error-Based Transfer Function Form	Page 100	Appendix A.4.2
Discrete-Time State-Space Form	Page 130	Appendix A.5
Discrete-Time Transfer Function Form	Page 134	Appendix A.6
Discrete-Time Dual-Feedback Transfer Function Form	Page 138	Appendix A.7
Discrete-Time Error-Based ADRC (all forms)	Page 143	Appendix A.8

Is This the End?

We are now in the last numbered chapter of the book, but there are still essential things coming. Appendix A repeats in a compact form all of the relevant equations required to implement and tune all the ADRC variants we have considered. Appendix B is a handy overview of various ADRC implementations in MATLAB/Simulink. It also details our proposed "Linear ADRC Blockset," which is an add-on helping to quickly deploy ADRC in the variants covered in the book. Therefore, we hope you will frequently return to this book, either to its main covered material (Parts I and II), to consult for the sought information regarding various aspects of ADRC, or to the Appendices, to use it as a reference.

From the perspective of ADRC, it is not the end either. Quite the contrary, from the scientific development side, many interesting topics are being investigated—some of which we briefly covered in the "look beyond" part of Chap. 7. From the practical side, we now expect its larger adoption in the real world for all the reasons associated with the ADRC area reaching its technological maturity phase.

As for the future of ADRC in a broader sense, it is expected that over the next decade, the biggest generators of real-time data will be devices that sense and control the physical world. Such an amount of data requires a rapprochement of the control area. Control theory has been traditionally and firmly rooted in model-based design. However, the availability and scale of emerging data (both temporal and spatial) will require us to rethink the foundations of our discipline. With its inherent large independence from a precise description of the controlled dynamical system, a data-driven methodology like ADRC could play a meaningful role in that endeavor.

Coming back to the question: Is this the end? For the regular part of the book, yes, it is. But we hope this will also be a start: for you putting ADRC into practice.

12 Postlude: A Look Ahead

Open Access This chapter is licensed under the terms of the Creative Commons Attribution 4.0 International License (http://creativecommons.org/licenses/by/4.0/), which permits use, sharing, adaptation, distribution and reproduction in any medium or format, as long as you give appropriate credit to the original author(s) and the source, provide a link to the Creative Commons license and indicate if changes were made.

The images or other third party material in this chapter are included in the chapter's Creative Commons license, unless indicated otherwise in a credit line to the material. If material is not included in the chapter's Creative Commons license and your intended use is not permitted by statutory regulation or exceeds the permitted use, you will need to obtain permission directly from the copyright holder.

Correction to: Active Disturbance Rejection Control

Correction to:
G. Herbst, R. Madonski, *Active Disturbance Rejection Control,*
Control Engineering, https://doi.org/10.1007/978-3-031-72687-3

Owing to an unfortunate oversight on the part of the publisher, the following errors were unintentionally introduced into the book titled "Active Disturbance Rejection Control" after the author had signed off on the book proof.

- In the table of contents "Part II Going Practical" is supposed to appear on page xii.
- In page 39, in Equation (3.20), a missing omega symbol on the right hand side.
- Extraneous white space in the recipe on page 131.
- The quality of the figures in Chapters 10 and 11 as well as Appendix B was diminished.
- Section numbering in Appendix B is incorrect.

The updated version of this book can be found at
https://doi.org/10.1007/978-3-031-72687-3

Open Access This chapter is licensed under the terms of the Creative Commons Attribution 4.0 International License (http://creativecommons.org/licenses/by/4.0/), which permits use, sharing, adaptation, distribution and reproduction in any medium or format, as long as you give appropriate credit to the original author(s) and the source, provide a link to the Creative Commons license and indicate if changes were made.

The images or other third party material in this chapter are included in the chapter's Creative Commons license, unless indicated otherwise in a credit line to the material. If material is not included in the chapter's Creative Commons license and your intended use is not permitted by statutory regulation or exceeds the permitted use, you will need to obtain permission directly from the copyright holder.

Appendix A
Linear ADRC Cheat Sheet

This chapter is meant to serve as a comprehensive reference guide to linear ADRC, repeating and summarizing all relevant equations to implement and tune all variants of ADRC that have been introduced in this book, be it in state-space or transfer function form, continuous or discrete time. For the particularly relevant first- and second-order cases—where ADRC replaces PI and PID controllers, respectively—all required coefficients are concisely presented in tabular form.

A.1 Continuous-Time ADRC, State-Space Form

The archetype of linear ADRC is its continuous-time state-space form. As depicted in Fig. A.1.1, the controller block has one output, the control signal $u(t)$, and three inputs: reference $r(t)$, measured plant output $y(t)$, and the limited control signal $u_{\text{lim}}(t)$. This form allows the application of user-defined limitations (e.g., magnitude and/or rate constraints) on the control signal, represented by the placeholder "Limiter" block in Fig. A.1.1.

The control law for continuous-time ADRC in state space is

$$u(t) = \frac{1}{b_0} \cdot \left(k_1 \cdot r(t) - \left(\boldsymbol{k}^{\text{T}}\ 1\right) \cdot \hat{\boldsymbol{x}}(t)\right) \tag{A.1.1}$$

with $\boldsymbol{k}^{\text{T}} = \begin{pmatrix} k_1 & \cdots & k_N \end{pmatrix}$ and $\hat{\boldsymbol{x}} = \begin{pmatrix} \hat{x}_1 & \cdots & \hat{x}_{N+1} \end{pmatrix}^{\text{T}}$,

while the observer dynamics are described by the following equation:

$$\dot{\hat{\boldsymbol{x}}}(t) = \boldsymbol{A} \cdot \hat{\boldsymbol{x}}(t) + \boldsymbol{b} \cdot u_{\text{lim}}(t) + \boldsymbol{l} \cdot \left(y(t) - \boldsymbol{c}^{\text{T}} \cdot \hat{\boldsymbol{x}}(t)\right) \quad \text{with} \quad \boldsymbol{l} = \begin{pmatrix} l_1 & \cdots & l_{N+1} \end{pmatrix}. \tag{A.1.2}$$

Note again that the controller output is $u(t)$, whereas the observer consumes $u_{\text{lim}}(t)$, the output of an arbitrary user-defined control signal limitation block. Controller and observer gains $\boldsymbol{k}^{\text{T}}$ and $\boldsymbol{l}$ obtained from bandwidth parameterization are given in Table A.1.1; matrix $\boldsymbol{A}$ and vectors $\boldsymbol{b}$, $\boldsymbol{c}^{\text{T}}$ of the observer in Table A.1.2.

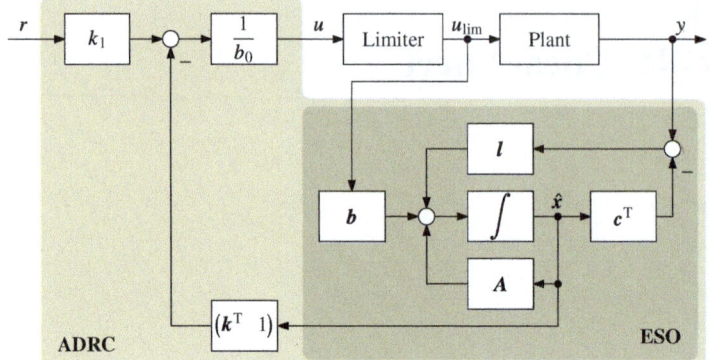

Fig. A.1.1 Continuous-time state-space ADRC, with an external, user-defined control signal limiter

Table A.1.1 Continuous-time ADRC, state-space form: controller and observer gains obtained using bandwidth parameterization. Tuning parameters are ω_{CL} (desired closed-loop bandwidth) and k_{ESO} (observer bandwidth factor relative to the closed-loop dynamics)

Order	Controller gains	Observer gains
1st	$k_1 = \omega_{CL}$	$l_1 = 2k_{ESO}\omega_{CL}$
		$l_2 = k_{ESO}^2 \omega_{CL}^2$
2nd	$k_1 = \omega_{CL}^2$	$l_1 = 3k_{ESO}\omega_{CL}$
	$k_2 = 2\omega_{CL}$	$l_2 = 3k_{ESO}^2 \omega_{CL}^2$
		$l_3 = k_{ESO}^3 \omega_{CL}^3$
Nth	$k_i = \dfrac{N!}{(N-i+1)! \cdot (i-1)!} \cdot \omega_{CL}^{N-i+1}$ $\forall i = 1, \ldots, N$	$l_i = \dfrac{(N+1)!}{(N-i+1)! \cdot i!} \cdot (k_{ESO} \cdot \omega_{CL})^i$ $\forall i = 1, \ldots, N+1$

Table A.1.2 Continuous-time ADRC, state-space form: matrices and vectors A, b, c^T of the extended state observer (ESO)

Order	A	b	c^T
1st	$\begin{pmatrix} 0 & 1 \\ 0 & 0 \end{pmatrix}$	$\begin{pmatrix} b_0 \\ 0 \end{pmatrix}$	$(1\ 0)$
2nd	$\begin{pmatrix} 0 & 1 & 0 \\ 0 & 0 & 1 \\ 0 & 0 & 0 \end{pmatrix}$	$\begin{pmatrix} 0 \\ b_0 \\ 0 \end{pmatrix}$	$(1\ 0\ 0)$
Nth	$\begin{pmatrix} \mathbf{0}^{N \times 1} & \mathbf{I}^{N \times N} \\ 0 & \mathbf{0}^{1 \times N} \end{pmatrix}$	$\begin{pmatrix} \mathbf{0}^{(N-1) \times 1} \\ b_0 \\ 0 \end{pmatrix}$	$(1\ \mathbf{0}^{1 \times N})$

A.2 Continuous-Time ADRC, Transfer Function Form

A transfer function form of continuous-time ADRC can be given using two transfer functions, as shown in Fig. A.2.1: a feedback controller $C_{\text{FB}}(s)$ (integrator with lead/lag elements) and a non-realizable prefilter $C_{\text{PF}}^{\text{NR}}(s)$, resulting the following control law:

$$u(s) = C_{\text{FB}}(s) \cdot \left[C_{\text{PF}}^{\text{NR}}(s) \cdot r(s) - y(s) \right]. \quad (\text{A.2.1})$$

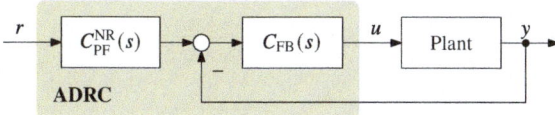

Fig. A.2.1 Control loop with a transfer function form of continuous-time ADRC, consisting of a feedback controller $C_{\text{FB}}(s)$ and a non-realizable prefilter $C_{\text{PF}}^{\text{NR}}(s)$

For the first- and second-order cases of ADRC, these transfer functions are given in Table A.2.1. The corresponding coefficients are presented in Tables A.2.2 and A.2.3 both in terms of bandwidth parameterization tuning parameters and in terms of controller/observer gains, to facilitate a transition from state space with arbitrary controller/observer tuning.

Table A.2.1 Continuous-time transfer function form of first- and second-order ADRCs. The corresponding α, β, γ coefficients and K_{I} are given in Tables A.2.2 and A.2.3, respectively

Order	Feedback controller	Reference signal prefilter
1st	$C_{\text{FB}}(s) = \dfrac{K_{\text{I}}}{s} \cdot \dfrac{1 + \beta_1 s}{1 + \alpha_1 s}$	$C_{\text{PF}}^{\text{NR}}(s) = \dfrac{1 + \gamma_1 s + \gamma_2 s^2}{1 + \beta_1 s}$
2nd	$C_{\text{FB}}(s) = \dfrac{K_{\text{I}}}{s} \cdot \dfrac{1 + \beta_1 s + \beta_2 s^2}{1 + \alpha_1 s + \alpha_2 s^2}$	$C_{\text{PF}}^{\text{NR}}(s) = \dfrac{1 + \gamma_1 s + \gamma_2 s^2 + \gamma_3 s^3}{1 + \beta_1 s + \beta_2 s^2}$

It is important to emphasize once more that this transfer function form is not realizable and therefore mostly meant for theoretical analyses. It cannot be implemented in practice, and even simulation systems might refuse to handle $C_{\text{PF}}^{\text{NR}}(s)$. A realizable transfer function form is given next in Appendix A.3.

Table A.2.2 Continuous-time ADRC: transfer function coefficients for the first-order case. For bandwidth parameterization, tuning parameters are b_0 (gain parameter of the plant model), ω_{CL} (desired closed-loop bandwidth), and k_{ESO} (relative bandwidth of observer compared to closed loop). "General terms" refer to an existing controller/observer tuning in state space, cf. Appendix A.1

Coeff.	General terms	Bandwidth parameterization
K_I	$\dfrac{1}{b_0} \cdot \dfrac{k_1 l_2}{k_1 + l_1}$	$\dfrac{1}{b_0} \cdot \dfrac{k_{ESO}^2 \omega_{CL}^2}{1 + 2k_{ESO}}$
α_1	$\dfrac{1}{k_1 + l_1}$	$\dfrac{1}{\omega_{CL} \cdot (1 + 2k_{ESO})}$
β_1	$\left(\dfrac{l_1}{l_2} + \dfrac{1}{k_1}\right)$	$\dfrac{2 + k_{ESO}}{k_{ESO}\omega_{CL}}$
γ_1	$\dfrac{l_1}{l_2}$	$\dfrac{2}{k_{ESO}\omega_{CL}}$
γ_2	$\dfrac{1}{l_2}$	$\dfrac{1}{k_{ESO}^2 \omega_{CL}^2}$

Table A.2.3 Continuous-time ADRC: transfer function coefficients for the second-order case. For bandwidth parameterization, tuning parameters are b_0 (gain parameter of the plant model), ω_{CL} (desired closed-loop bandwidth), and k_{ESO} (relative bandwidth of the observer compared the closed loop). "General terms" refer to an existing controller/observer tuning in state space, cf. Appendix A.1

Coeff.	General terms	Bandwidth parameterization
K_I	$\dfrac{1}{b_0} \cdot \dfrac{k_1 l_3}{k_1 + k_2 l_1 + l_2}$	$\dfrac{1}{b_0} \cdot \dfrac{k_{ESO}^3 \omega_{CL}^3}{1 + 6k_{ESO} + 3k_{ESO}^2}$
α_1	$\dfrac{k_2 + l_1}{k_1 + k_2 l_1 + l_2}$	$\dfrac{2 + 3k_{ESO}}{\omega_{CL} \cdot \left(1 + 6k_{ESO} + 3k_{ESO}^2\right)}$
α_2	$\dfrac{1}{k_1 + k_2 l_1 + l_2}$	$\dfrac{1}{\omega_{CL}^2 \cdot \left(1 + 6k_{ESO} + 3k_{ESO}^2\right)}$
β_1	$\left(\dfrac{l_2}{l_3} + \dfrac{k_2}{k_1}\right)$	$\dfrac{1}{\omega_{CL}} \cdot \left(\dfrac{3}{k_{ESO}} + 2\right)$
β_2	$\left(\dfrac{l_1}{l_3} + \dfrac{k_2}{k_1} \cdot \dfrac{l_2}{l_3} + \dfrac{1}{k_1}\right)$	$\dfrac{1}{\omega_{CL}^2} \cdot \left(\dfrac{3}{k_{ESO}^2} + \dfrac{6}{k_{ESO}} + 1\right)$
γ_1	$\dfrac{l_2}{l_3}$	$\dfrac{3}{k_{ESO}\omega_{CL}}$
γ_2	$\dfrac{l_1}{l_3}$	$\dfrac{3}{k_{ESO}^2 \omega_{CL}^2}$
γ_3	$\dfrac{1}{l_3}$	$\dfrac{1}{k_{ESO}^3 \omega_{CL}^3}$

A.3 Continuous-Time ADRC, Realizable Transfer Function Form

A realizable transfer function form of continuous-time ADRC can be found using three transfer functions, leading to the following control law visualized in Fig. A.3.1:

$$u(s) = C_{\text{FB}}(s) \cdot [C_{\text{PF}}(s) \cdot r(s) - y(s)] + C_{\text{FF}}(s) \cdot r(s). \quad (\text{A.3.1})$$

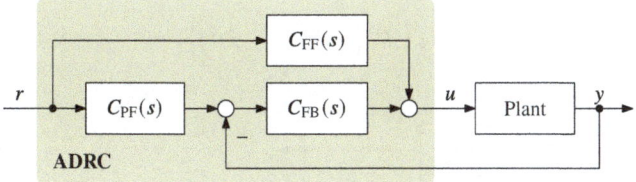

Fig. A.3.1 Continuous-time ADRC, realizable transfer function form consisting of feedback controller $C_{\text{FB}}(s)$, prefilter $C_{\text{PF}}(s)$, and a feedforward $C_{\text{FF}}(s)$

Compared to the non-realizable form of Appendix A.2, a new feedforward block $C_{\text{FF}}(s)$ with high-pass filter characteristics has been added, which makes the prefilter $C_{\text{PF}}(s)$ realizable. For the first- and second-order cases of ADRC, these transfer functions are given in Table A.3.1. Note that the corresponding coefficients are the same as for the non-realizable form, which means that the values from Tables A.2.2 and A.2.3 apply here, as well.

Table A.3.1 Continuous-time realizable transfer function form of first- and second-order ADRCs. The α, β, γ coefficients and K_I are given in Tables A.2.2 and A.2.3, respectively

Order	Feedback controller	Reference signal prefilter	Reference signal feedforward
1st	$C_{\text{FB}}(s) = \dfrac{K_\text{I}}{s} \cdot \dfrac{1 + \beta_1 s}{1 + \alpha_1 s}$	$C_{\text{PF}}(s) = \dfrac{1 + \gamma_1 s}{1 + \beta_1 s}$	$C_{\text{FF}}(s) = \dfrac{K_\text{I} \gamma_2 s}{1 + \alpha_1 s}$
2nd	$C_{\text{FB}}(s) = \dfrac{K_\text{I}}{s} \cdot \dfrac{1 + \beta_1 s + \beta_2 s^2}{1 + \alpha_1 s + \alpha_2 s^2}$	$C_{\text{PF}}(s) = \dfrac{1 + \gamma_1 s + \gamma_2 s^2}{1 + \beta_1 s + \beta_2 s^2}$	$C_{\text{FF}}(s) = \dfrac{K_\text{I} \gamma_3 s^2}{1 + \alpha_1 s + \alpha_2 s^2}$

A drawback of this transfer function form compared to the continuous-time state-space form is that a control signal limitation with windup protection cannot be incorporated here as easily as before, since the inner feedback path of the limited control signal output through the observer got eliminated in the process of deriving the feedback controller transfer function.

A.4 Continuous-Time Error-Based ADRC

A.4.1 State-Space Form

A block diagram of error-based ADRC in state space is shown in Fig. A.4.1. Matrix A and vectors b, c^T and controller and observer gains k^T and l are identical to the output-based variant from Appendix A.1. They can be found in Tables A.1.1 and A.1.1, respectively. Control law and observer dynamics are given by:

$$u(t) = \frac{1}{b_0} \cdot \left(k^T \ 1\right) \cdot \hat{x}(t), \qquad (A.4.1)$$

$$\dot{\hat{x}}(t) = A \cdot \hat{x}(t) - b \cdot u_{\text{lim}}(t) + l \cdot \left(e(t) - c^T \cdot \hat{x}(t)\right). \qquad (A.4.2)$$

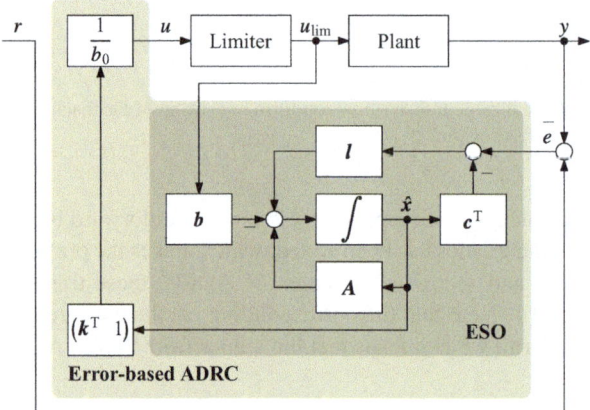

Fig. A.4.1 Continuous-time error-based ADRC, with external control signal limitation

A.4.2 Transfer Function Form

The transfer function form of error-based ADRC in (A.4.3) consists only of a feedback controller, cf. Fig. A.4.2, which is identical to $C_{\text{FB}}(s)$ of output-based ADRC in Appendix A.2. For the first- and second-order cases, transfer functions can accordingly be taken from Table A.2.1, along with their coefficients in Tables A.2.2 and A.2.3. Output-based ADRC can therefore be turned into error-based ADRC by leaving out the prefilter (and, in the realizable case, feedforward) transfer functions.

$$u(s) = C_{\text{PF}}(s) \cdot e(s) \qquad (A.4.3)$$

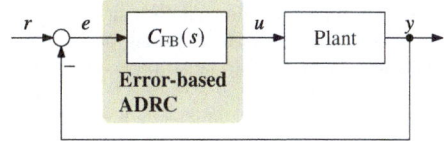

Fig. A.4.2 Continuous-time error-based ADRC, transfer function form

A.5 Discrete-Time ADRC, State-Space Form

A discrete-time form of ADRC in state space is shown in Fig. A.5.1, resulting from a zero-order hold discretization and the preferable use of the so-called *current observer* approach. The control law is

$$u(k) = \frac{1}{b_0} \cdot \left(k_1 \cdot r(k) - \begin{pmatrix} \boldsymbol{k}^T & 1 \end{pmatrix} \cdot \hat{\boldsymbol{x}}(k)\right) \quad \text{with} \quad \boldsymbol{k}^T = \begin{pmatrix} k_1 & \cdots & k_N \end{pmatrix}, \quad \text{(A.5.1)}$$

while the observer is being governed by an equation that can be formulated in a compact form with abbreviations $\boldsymbol{A}_{\text{ESO}}$ and $\boldsymbol{b}_{\text{ESO}}$:

$$\hat{\boldsymbol{x}}(k) = \boldsymbol{A}_{\text{ESO}} \cdot \hat{\boldsymbol{x}}(k-1) + \boldsymbol{b}_{\text{ESO}} \cdot u_{\text{lim}}(k-1) + \boldsymbol{l} \cdot y(k) \quad \text{with} \quad \boldsymbol{l} = \begin{pmatrix} l_1 & \cdots & l_{N+1} \end{pmatrix}^T. \quad \text{(A.5.2)}$$

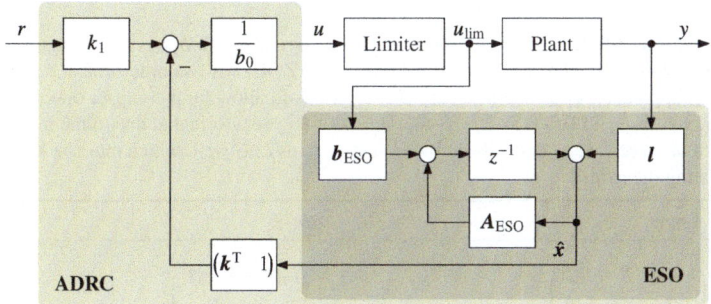

Fig. A.5.1 Discrete-time ADRC, state-space form, with an external control signal limiter

The controller and observer gains $\boldsymbol{k}^T$ and $\boldsymbol{l}$ obtained from bandwidth parameterization can be found in Table A.5.1. The matrix $\boldsymbol{A}_{\text{ESO}}$ and vector $\boldsymbol{b}_{\text{ESO}}$ for implementing the observer equations are given in full detail in Table A.5.2.

As in the continuous-time case in Appendix A.1, it should be noted that although $u(k)$ is the output of the control law, it is $u_{\text{lim}}(k)$ that is being fed into the observer. $u_{\text{lim}}(k)$ is the output of an arbitrary user-defined control signal limitation block denoted as "Limiter" in Fig. A.5.1. A possible example enforcing magnitude and discrete-time rate constraints is shown in Fig. 9.3 on page 152.

It is an important advantage of ADRC that using $u_{\text{lim}}(k)$ instead of $u(k)$ as an input to the observer is the only necessary measure to prevent possible windup issues in the controller—besides using a limiter block in the first place, of course.

Table A.5.1 Discrete-time ADRC, state-space form: controller and observer gains obtained from bandwidth parameterization. Tuning parameters are T_s (sample time), ω_{CL} (desired closed-loop bandwidth), and k_{ESO} (relative observer bandwidth). From these, the z-domain pole locations $z_{CL} = e^{-\omega_{CL} T_s}$ and $z_{ESO} = e^{-k_{ESO} \omega_{CL} T_s}$ are computed to be put into the given equations. Note that in most practical cases (sufficiently high sampling frequencies), the controller gains are basically the same for continuous and discrete time, cf. Table A.1.1. The given approximations are valid if $\omega_{CL} T_s \ll 1$ holds

Order	Controller gains	Observer gains
1st	$k_1 = \dfrac{1 - z_{CL}}{T_s} \approx \omega_{CL}$	$l_1 = 1 - z_{ESO}^2$ $l_2 = \dfrac{1}{T_s} \cdot (1 - z_{ESO})^2$
2nd	$k_1 = \dfrac{(1 - z_{CL})^2}{T_s^2} \approx \omega_{CL}^2$ $k_2 = \dfrac{4 - (1 + z_{CL})^2}{2 T_s} \approx 2\omega_{CL}$	$l_1 = 1 - z_{ESO}^3$ $l_2 = \dfrac{3}{2 T_s} \cdot (1 - z_{ESO})^2 \cdot (1 + z_{ESO})$ $l_3 = \dfrac{1}{T_s^2} \cdot (1 - z_{ESO})^3$
Nth [†]	Pole placement, i.e., solution of $\det\left(z\boldsymbol{I} - \left(\boldsymbol{A}_{VP,d} - \boldsymbol{b}_{VP,d}\boldsymbol{k}^T\right)\right) = (z - z_{CL})^N$	Pole placement, i.e., solution of $\det(z\boldsymbol{I} - \boldsymbol{A}_{ESO}) = (z - z_{ESO})^{N+1}$

[†] The matrix $\boldsymbol{A}_{ESO}$ definition is given in Table A.5.2 and $\boldsymbol{A}_{VP,d}$ and $\boldsymbol{b}_{VP,d}$ in (8.12)

Table A.5.2 Discrete-time ADRC, state-space form: matrix $\boldsymbol{A}_{ESO}$ and vector $\boldsymbol{b}_{ESO}$ of the extended state observer (ESO), based on zero-order hold discretization (ZOH, with sample time T_s) and the *current observer* approach. $\boldsymbol{A}_{ESO}$ and $\boldsymbol{b}_{ESO}$ are abbreviations to allow for a compact observer implementation $\hat{\boldsymbol{x}}(k) = \boldsymbol{A}_{ESO} \cdot \hat{\boldsymbol{x}}(k-1) + \boldsymbol{b}_{ESO} \cdot u_{\lim}(k-1) + \boldsymbol{l} \cdot y(k)$. Note that due to the inner feedback loop of an observer, $\boldsymbol{A}_{ESO}$ depends on the observer gains $\boldsymbol{l}$, as well, which must be kept in mind when updating the gains $\boldsymbol{l}$

Order	$\boldsymbol{A}_{ESO}$	$\boldsymbol{b}_{ESO}$
1st	$\begin{pmatrix} 1 - l_1 & T_s - l_1 T_s \\ -l_2 & 1 - l_2 T_s \end{pmatrix}$	$\begin{pmatrix} b_0 T_s - l_1 b_0 T_s \\ -l_2 b_0 T_s \end{pmatrix}$
2nd	$\begin{pmatrix} 1 - l_1 & T_s - l_1 T_s & \tfrac{1}{2} T_s^2 - \tfrac{1}{2} l_1 T_s^2 \\ -l_2 & 1 - l_2 T_s & T_s - \tfrac{1}{2} l_2 T_s^2 \\ -l_3 & -l_3 T_s & 1 - \tfrac{1}{2} l_3 T_s^2 \end{pmatrix}$	$\begin{pmatrix} \tfrac{1}{2} b_0 T_s^2 - \tfrac{1}{2} l_1 b_0 T_s^2 \\ b_0 T_s - \tfrac{1}{2} l_2 b_0 T_s^2 \\ -\tfrac{1}{2} l_3 b_0 T_s^2 \end{pmatrix}$
Nth [†]	$\boldsymbol{A}_d - \boldsymbol{l} \cdot \boldsymbol{c}_d^T \cdot \boldsymbol{A}_d$ with $\boldsymbol{A}_d = \boldsymbol{I} + \sum_{i=1}^{\infty} \dfrac{\boldsymbol{A}^i T_s^i}{i!}$, $\boldsymbol{b}_d = \left(\sum_{i=1}^{\infty} \dfrac{\boldsymbol{A}^{i-1} T_s^i}{i!}\right) \cdot \boldsymbol{b}$, $\boldsymbol{c}_d^T = \boldsymbol{c}^T$	$\boldsymbol{b}_d - \boldsymbol{l} \cdot \boldsymbol{c}_d^T \cdot \boldsymbol{b}_d$

[†] Definitions of the continuous-time matrices and vectors $\boldsymbol{A}, \boldsymbol{b}, \boldsymbol{c}^T$ are given in Table A.1.2

A.6 Discrete-Time ADRC, Transfer Function Form

In contrast to the continuous-time case, there are no realizability issues for a discrete-time transfer function form of ADRC. A variant using two transfer functions—a feedback controller $C_{\text{FB}}(z)$ and a prefilter $C_{\text{PF}}(z)$—is depicted in Fig. A.6.1, with the following control law:

$$u(z) = C_{\text{FB}}(z) \cdot [C_{\text{PF}}(z) \cdot r(z) - y(z)]. \tag{A.6.1}$$

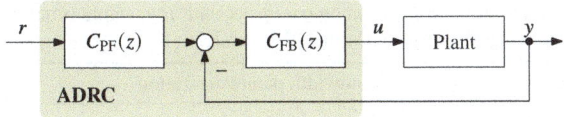

Fig. A.6.1 Discrete-time ADRC, transfer function form, consisting of a feedback controller $C_{\text{FB}}(z)$ and a prefilter $C_{\text{PF}}(z)$

The transfer functions $C_{\text{FB}}(z)$ and $C_{\text{PF}}(z)$ are given in Table A.6.1 and the corresponding coefficients for the first- and second-order cases in Tables A.6.2 and A.6.3, respectively.

Table A.6.1 Transfer functions of the discrete-time ADRC form from Fig. A.6.1. The coefficients α, β, γ are given in Tables A.6.2 and A.6.3 for the first- and second order cases

Order	Feedback controller	Reference signal prefilter
1st	$C_{\text{FB}}(z) = \dfrac{\beta_0 + \beta_1 z^{-1}}{1 + \alpha_1 z^{-1}} \cdot \dfrac{1}{1 - z^{-1}}$	$C_{\text{PF}}(z) = \dfrac{\frac{\gamma_0}{\beta_0} + \frac{\gamma_1}{\beta_0} z^{-1} + \frac{\gamma_2}{\beta_0} z^{-2}}{1 + \frac{\beta_1}{\beta_0} z^{-1}}$
2nd	$C_{\text{FB}}(z) = \dfrac{\beta_0 + \beta_1 z^{-1} + \beta_2 z^{-2}}{1 + \alpha_1 z^{-1} + \alpha_2 z^{-2}} \cdot \dfrac{1}{1 - z^{-1}}$	$C_{\text{PF}}(z) = \dfrac{\frac{\gamma_0}{\beta_0} + \frac{\gamma_1}{\beta_0} z^{-1} + \frac{\gamma_2}{\beta_0} z^{-2} + \frac{\gamma_3}{\beta_0} z^{-3}}{1 + \frac{\beta_1}{\beta_0} z^{-1} + \frac{\beta_2}{\beta_0} z^{-2}}$
Nth	$C_{\text{FB}}(z) = \dfrac{\sum_{i=0}^{n} \beta_i z^{-i}}{1 + \sum_{i=1}^{n} \alpha_i z^{-i}} \cdot \dfrac{1}{1 - z^{-1}}$	$C_{\text{PF}}(z) = \dfrac{\frac{1}{\beta_0} \sum_{i=0}^{n+1} \gamma_i z^{-i}}{1 + \frac{1}{\beta_0} \sum_{i=1}^{n} \beta_i z^{-i}}$

It is immediately obvious from Table A.6.1 that the feedback controller $C_{\text{FB}}(z)$ includes a discrete-time integrator pole, which is presented in a factored-out form for several reasons:

- The structural similarity to the continuous-time form becomes more apparent, as the integrator is factored out in Tables A.2.1 and A.3.1, as well.

- This form requires one coefficient less compared to a form where the integrator pole is incorporated in the denominator polynomial, therefore saving one multiplication at runtime.
- It allows incorporating at least a basic form of anti-windup mechanism by using a saturated integrator when implementing the integrator pole.

Table A.6.2 Discrete-time ADRC: transfer function coefficients for the first-order case. "General terms" refer to an existing discrete-time state-space controller/observer tuning, cf. Appendix A.5. Bandwidth parameterization uses discrete-time pole placement at $z_{\text{CL}} = e^{-\omega_{\text{CL}} T_s}$ (closed loop) and $z_{\text{ESO}} = e^{-k_{\text{ESO}} \omega_{\text{CL}} T_s}$ (observer), based on the tuning parameters ω_{CL} (desired closed-loop bandwidth) and k_{ESO} (relative observer bandwidth). Common parameters are T_s (sample time of the discrete-time implementation) and b_0 (gain parameter of the plant model)

Coeff.	General terms	Bandwidth parameterization
α_1	$(T_s k_1 - 1) \cdot (1 - l_1)$	$-z_{\text{CL}} z_{\text{ESO}}^2$
β_0	$\dfrac{1}{b_0} \cdot (k_1 l_1 + l_2)$	$\dfrac{1}{b_0 T_s} \cdot \left(z_{\text{CL}} z_{\text{ESO}}^2 - 2 z_{\text{ESO}} - z_{\text{CL}} + 2 \right)$
β_1	$\dfrac{1}{b_0} \cdot (T_s k_1 l_2 - k_1 l_1 - l_2)$	$\dfrac{1}{b_0 T_s} \cdot \left(2 z_{\text{CL}} z_{\text{ESO}} - 2 z_{\text{CL}} z_{\text{ESO}}^2 + z_{\text{ESO}}^2 - 1 \right)$
γ_0	$\dfrac{k_1}{b_0}$	$\dfrac{1 - z_{\text{CL}}}{b_0 T_s}$
γ_1	$\dfrac{k_1}{b_0} \cdot (T_s l_2 + l_1 - 2)$	$\dfrac{-2 z_{\text{ESO}} \cdot (1 - z_{\text{CL}})}{b_0 T_s}$
γ_2	$\dfrac{k_1}{b_0} \cdot (1 - l_1)$	$\dfrac{z_{\text{ESO}}^2 \cdot (1 - z_{\text{CL}})}{b_0 T_s}$

This structure with prefilter and feedback controller is common in control engineering. It allows to better compare ADRC against other discrete-time controllers or even to implement ADRC, e.g., using existing software blocks corresponding to these transfer functions. It also requires fewer multiplications than a state-space implementation. Yet it has to be noted that the anti-windup behavior differs from the more powerful state-space form as $u_{\text{lim}}(k)$ is being fed to the observer in the latter—a feedback path that gets removed when deriving the transfer functions. A variant maintaining the exact anti-windup behavior of the discrete-time state-space version while providing an even lower computational footprint than these two transfer functions is possible, however, and presented next in Appendix A.7.

A Linear ADRC Cheat Sheet

Table A.6.3 Discrete-time ADRC: transfer function coefficients for the second-order case. "General terms" refer to an existing discrete-time state-space controller/observer tuning, cf. Appendix A.5. Bandwidth parameterization uses discrete-time pole placement at $z_{\text{CL}} = e^{-\omega_{\text{CL}} T_s}$ (closed loop) and $z_{\text{ESO}} = e^{-k_{\text{ESO}} \omega_{\text{CL}} T_s}$ (observer), based on the tuning parameters ω_{CL} (desired closed-loop bandwidth) and k_{ESO} (relative observer bandwidth). Common parameters are T_s (sample time of the discrete-time implementation) and b_0 (gain parameter of the plant model)

Coeff.	General terms	Bandwidth parameterization
α_1	$\dfrac{T_s^2}{2} \cdot (k_1 - k_1 l_1 - k_2 l_2) + T_s k_2 + T_s l_2 + l_1 - 2$	$-\dfrac{1}{8}(1+z_{\text{CL}})^2(1+z_{\text{ESO}})^3 + z_{\text{CL}}^2 z_{\text{ESO}}^3 + 1$
α_2	$\left(\dfrac{T_s^2 k_1}{2} - T_s k_2 + 1\right) \cdot (1 - l_1)$	$z_{\text{CL}}^2 z_{\text{ESO}}^3$
β_0	$\dfrac{1}{b_0} \cdot [k_1 l_1 + k_2 l_2 + l_3]$	$\dfrac{1}{b_0 T_s^2} \cdot \left[\dfrac{1}{4}(1+z_{\text{CL}})^2(1+z_{\text{ESO}})^3 - 2\left(z_{\text{CL}}^2 z_{\text{ESO}}^3 + 2z_{\text{CL}} + 3z_{\text{ESO}} - 2\right) \right]$
β_1	$\dfrac{1}{b_0} \cdot \left[\dfrac{T_s^2 k_1 l_3}{2} + T_s k_1 l_2 + T_s k_2 l_3 - 2(k_1 l_1 + k_2 l_2 + l_3) \right]$	$\dfrac{1}{b_0 T_s^2} \cdot \left[-(1+z_{\text{CL}})^2(1+z_{\text{ESO}})^3 + 2(1+z_{\text{CL}})^2 + 6\left(z_{\text{CL}}^2 z_{\text{ESO}}^3 + 2z_{\text{CL}} z_{\text{ESO}} + z_{\text{ESO}}^2 - 1\right) \right]$
β_2	$\dfrac{1}{b_0} \cdot \left[\dfrac{T_s^2 k_1 l_3}{2} - T_s k_1 l_2 - T_s k_2 l_3 + (k_1 l_1 + k_2 l_2 + l_3) \right]$	$\dfrac{1}{b_0 T_s^2} \cdot \left[-\dfrac{1}{4}(1+z_{\text{CL}})^2(1+z_{\text{ESO}})^3 + 2\left(-2z_{\text{CL}}^2 z_{\text{ESO}}^3 + 3z_{\text{CL}}^2 z_{\text{ESO}}^2 + 2z_{\text{CL}} z_{\text{ESO}}^3 + 1\right) \right]$
γ_0	$\dfrac{k_1}{b_0}$	$\dfrac{(1-z_{\text{CL}})^2}{b_0 T_s^2}$
γ_1	$\dfrac{k_1}{b_0} \cdot \left(\dfrac{T_s^2 l_3}{2} + T_s l_2 + l_1 - 3 \right)$	$\dfrac{-3z_{\text{ESO}}(1-z_{\text{CL}})^2}{b_0 T_s^2}$
γ_2	$\dfrac{k_1}{b_0} \cdot \left(\dfrac{T_s^2 l_3}{2} - T_s l_2 - 2l_1 + 3 \right)$	$\dfrac{3z_{\text{ESO}}^2(1-z_{\text{CL}})^2}{b_0 T_s^2}$
γ_3	$\dfrac{k_1}{b_0} \cdot (l_1 - 1)$	$\dfrac{-z_{\text{ESO}}^3(1-z_{\text{CL}})^2}{b_0 T_s^2}$

A.7 Discrete-Time ADRC, Dual-Feedback Transfer Function Form

A discrete-time transfer function form combining the exact behavior from the state-space version with the lower computational footprint of transfer functions is shown in Fig. A.7.1. It consists of two transfer functions placed in the feedback path and a single gain factor acting on $r(k)$, resulting in the control law:

$$u(z) = \frac{k_1}{b_0} \cdot r(z) - C_{\text{FBy}}(z) \cdot y(z) + C_{\text{FBu}}(z) \cdot u_{\lim}(z). \tag{A.7.1}$$

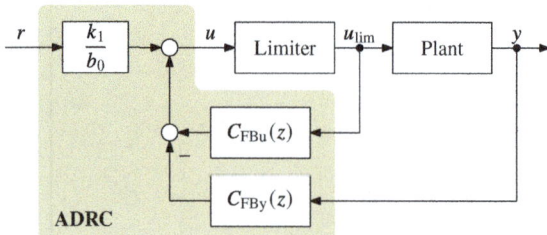

Fig. A.7.1 Discrete-time ADRC, dual-feedback transfer function form

In contrast to the transfer function approach of Appendix A.6, the explicit feedback of the limited control signal $u_{\lim}(k)$ is preserved here, allowing users to freely choose a control signal limitation block as in the state-space case of Appendix A.5.

The transfer functions $C_{\text{FBy}}(z)$ and $C_{\text{FBu}}(z)$ are given in Table A.7.1 and the required coefficients for the first- and second-order cases in Tables A.7.2 and A.7.3. Making use of superposition and a transposed direct form II, their computational footprint can be reduced to a minimum, shown in detail in Fig. A.7.2.

Table A.7.1 Transfer functions for the dual-feedback discrete-time ADRC form. Coefficients α, β, γ are given in Tables A.7.2 and A.7.3 for the first- and second-order cases

Order	Plant output feedback	Control signal feedback
1st	$C_{\text{FBy}}(z) = \dfrac{\beta_0 + \beta_1 z^{-1}}{1 + \alpha_1 z^{-1} + \alpha_2 z^{-2}}$	$C_{\text{FBu}}(z) = z^{-1} \cdot \dfrac{\gamma_0 + \gamma_1 z^{-1}}{1 + \alpha_1 z^{-1} + \alpha_2 z^{-2}}$
2nd	$C_{\text{FBy}}(z) = \dfrac{\beta_0 + \beta_1 z^{-1} + \beta_2 z^{-2}}{1 + \alpha_1 z^{-1} + \alpha_2 z^{-2} + \alpha_3 z^{-3}}$	$C_{\text{FBu}}(z) = z^{-1} \cdot \dfrac{\gamma_0 + \gamma_1 z^{-1} + \gamma_2 z^{-2}}{1 + \alpha_1 z^{-1} + \alpha_2 z^{-2} + \alpha_3 z^{-3}}$
Nth [†]	$C_{\text{FBy}}(z) = \dfrac{1}{b_0} \cdot (\boldsymbol{k}^{\text{T}} \ 1) \cdot \left(\boldsymbol{I} - z^{-1} \boldsymbol{A}_{\text{ESO}}\right)^{-1} \cdot \boldsymbol{l}$	$C_{\text{FBu}}(z) = -\dfrac{z^{-1}}{b_0} \cdot (\boldsymbol{k}^{\text{T}} \ 1) \cdot \left(\boldsymbol{I} - z^{-1} \boldsymbol{A}_{\text{ESO}}\right)^{-1} \cdot \boldsymbol{b}_{\text{ESO}}$

[†] Matrix $\boldsymbol{A}_{\text{ESO}}$ and vectors $\boldsymbol{b}_{\text{ESO}}, \boldsymbol{k}^{\text{T}}, \boldsymbol{l}$ are given in Tables A.5.1 and A.5.2.

A Linear ADRC Cheat Sheet

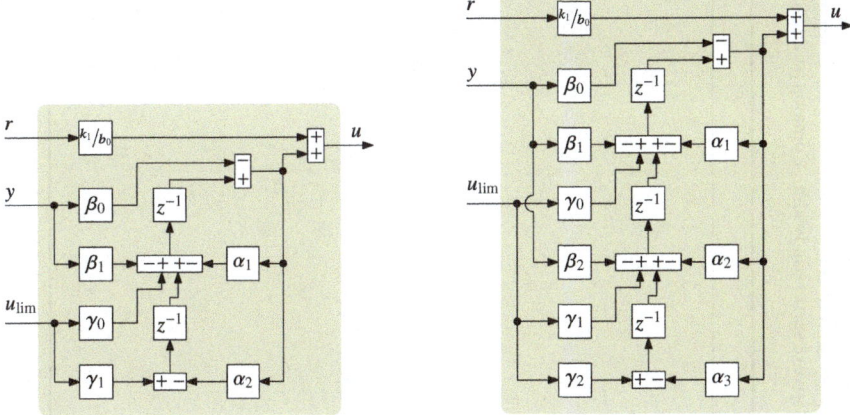

Fig. A.7.2 Minimum-footprint form of first- and second-order discrete-time ADRCs based on the dual-feedback transfer function form from Fig. A.7.1. As evident from Table A.7.1, $C_{\text{FBy}}(z)$ and $C_{\text{FBu}}(z)$ share the same denominator. The transposed direct form II implementations shown here therefore make use of superposition, resulting in a minimized number of multiplications and storage variables. Coefficients are given in Tables A.7.2 and A.7.3

Table A.7.2 Discrete-time ADRC, dual-feedback transfer function form: coefficients for the first-order case. "General terms" refer to an existing discrete-time state-space controller/observer tuning, cf. Appendix A.5. Bandwidth parameterization uses discrete-time pole placement at $z_{\text{CL}} = e^{-\omega_{\text{CL}} T_s}$ (closed loop) and $z_{\text{ESO}} = e^{-k_{\text{ESO}} \omega_{\text{CL}} T_s}$ (observer), based on the tuning parameters ω_{CL} (desired closed-loop bandwidth) and k_{ESO} (relative observer bandwidth). Common parameters are T_s (sample time of the discrete-time implementation) and b_0 (gain parameter of the plant model)

Coeff.	General terms	Bandwidth parameterization
α_1	$l_1 + T_s l_2 - 2$	$-2z_{\text{ESO}}$
α_2	$1 - l_1$	z_{ESO}^2
β_0	$\dfrac{1}{b_0} \cdot (k_1 l_1 + l_2)$	$\dfrac{1}{b_0 T_s} \cdot \left(z_{\text{CL}} z_{\text{ESO}}^2 - 2z_{\text{ESO}} - z_{\text{CL}} + 2 \right)$
β_1	$\dfrac{1}{b_0} \cdot (T_s k_1 l_2 - k_1 l_1 - l_2)$	$\dfrac{1}{b_0 T_s} \cdot \left(2z_{\text{CL}} z_{\text{ESO}} - 2z_{\text{CL}} z_{\text{ESO}}^2 + z_{\text{ESO}}^2 - 1 \right)$
γ_0	$T_s \cdot (-k_1 + k_1 l_1 + l_2)$	$z_{\text{CL}} z_{\text{ESO}}^2 - 2z_{\text{ESO}} + 1$
γ_1	$T_s k_1 \cdot (1 - l_1)$	$z_{\text{ESO}}^2 - z_{\text{CL}} z_{\text{ESO}}^2$
$\dfrac{k_1}{b_0}$	$\dfrac{k_1}{b_0}$	$\dfrac{1 - z_{\text{CL}}}{b_0 T_s}$

Table A.7.3 Discrete-time ADRC, dual-feedback transfer function form: coefficients for the second-order case. "General terms" refer to an existing discrete-time state-space controller/observer tuning, cf. Appendix A.5. Bandwidth parameterization uses discrete-time pole placement at $z_{\text{CL}} = e^{-\omega_{\text{CL}} T_s}$ (closed loop) and $z_{\text{ESO}} = e^{-k_{\text{ESO}} \omega_{\text{CL}} T_s}$ (observer), based on the tuning parameters ω_{CL} (desired closed-loop bandwidth) and k_{ESO} (relative observer bandwidth). Common parameters are T_s (sample time of the discrete-time implementation) and b_0 (gain parameter of the plant model)

Coeff.	General terms	Bandwidth parameterization
α_1	$l_1 + T_s l_2 + \dfrac{T_s^2 l_3}{2} - 3$	$-3 z_{\text{ESO}}$
α_2	$-2 l_1 - T_s l_2 + \dfrac{T_s^2 l_3}{2} + 3$	$3 z_{\text{ESO}}^2$
α_3	$l_1 - 1$	$-z_{\text{ESO}}^3$
β_0	$\dfrac{1}{b_0} \cdot [k_1 l_1 + k_2 l_2 + l_3]$	$\dfrac{1}{b_0 T_s^2} \cdot \left[\dfrac{1}{4}(1+z_{\text{CL}})^2 (1+z_{\text{ESO}})^3 - 2\left(z_{\text{CL}}^2 z_{\text{ESO}}^3 + 2 z_{\text{CL}} + 3 z_{\text{ESO}} - 2\right) \right]$
β_1	$\dfrac{1}{b_0} \cdot \left[\dfrac{T_s^2 k_1 l_3}{2} + T_s^2 k_1 l_2 + T_s k_2 l_3 - 2(k_1 l_1 + k_2 l_2 + l_3) \right]$	$\dfrac{1}{b_0 T_s^2} \cdot \left[-(1+z_{\text{CL}})^2 (1+z_{\text{ESO}})^3 + 2(1+z_{\text{CL}})^2 + 6\left(z_{\text{CL}}^2 z_{\text{ESO}}^3 + 2 z_{\text{CL}} z_{\text{ESO}} + z_{\text{ESO}}^2 - 1\right) \right]$
β_2	$\dfrac{1}{b_0} \cdot \left[\dfrac{T_s^2 k_1 l_3}{2} - T_s k_1 l_2 - T_s k_2 l_3 + (k_1 l_1 + k_2 l_2 + l_3) \right]$	$\dfrac{1}{b_0 T_s^2} \cdot \left[-\dfrac{1}{4}(1+z_{\text{CL}})^2 (1+z_{\text{ESO}})^3 + 2\left(-2 z_{\text{CL}}^2 z_{\text{ESO}}^3 + 3 z_{\text{CL}}^2 z_{\text{ESO}}^2 + 2 z_{\text{CL}} z_{\text{ESO}}^3 + 1\right) \right]$
γ_0	$\dfrac{T_s}{2} \cdot [-T_s k_1 + T_s k_1 l_1 - 2 k_2 + T_s k_2 l_2 + T_s l_3]$	$\dfrac{1}{8}(1+z_{\text{CL}})^2 (1+z_{\text{ESO}})^3 - z_{\text{ESO}}(z_{\text{CL}}^2 z_{\text{ESO}}^2 + 3)$
γ_1	$\dfrac{T_s}{2} \cdot [4 k_2 - 2 k_2 l_1 - T_s k_2 l_2 + T_s l_3]$	$-\dfrac{1}{8}(1+z_{\text{CL}})^2 (1+z_{\text{ESO}})^3 + 3 z_{\text{ESO}}^2 + 1$
γ_2	$\dfrac{T_s}{2} \cdot [(T_s k_1 - 2 k_2) \cdot (1 - l_1)]$	$z_{\text{ESO}}^3 \left(z_{\text{CL}}^2 - 1\right)$
$\dfrac{k_1}{b_0}$	$\dfrac{k_1}{b_0}$	$\dfrac{(1 - z_{\text{CL}})^2}{b_0 T_s^2}$

A.8 Discrete-Time Error-Based ADRC

A.8.1 State-Space Form

As the counterpart of Appendix A.5, the discrete-time state-space form of error-based ADRC is shown in Fig. A.8.1. The equations are

$$u(k) = \frac{1}{b_0} \cdot \begin{pmatrix} \boldsymbol{k}^T & 1 \end{pmatrix} \cdot \hat{\boldsymbol{x}}(k), \tag{A.8.1}$$

$$\hat{\boldsymbol{x}}(k) = \boldsymbol{A}_{\text{ESO}} \cdot \hat{\boldsymbol{x}}(k-1) - \boldsymbol{b}_{\text{ESO}} \cdot u_{\text{lim}}(k-1) + \boldsymbol{l} \cdot e(k). \tag{A.8.2}$$

The observer matrix $\boldsymbol{A}_{\text{ESO}}$ and vector $\boldsymbol{b}_{\text{ESO}}$ are identical to discrete-time output-based ADRC and can be found in Table A.5.2. Tuning remains unchanged, as well, and controller and observer gains $\boldsymbol{k}^T$ and $\boldsymbol{l}$ are therefore available in Table A.5.1.

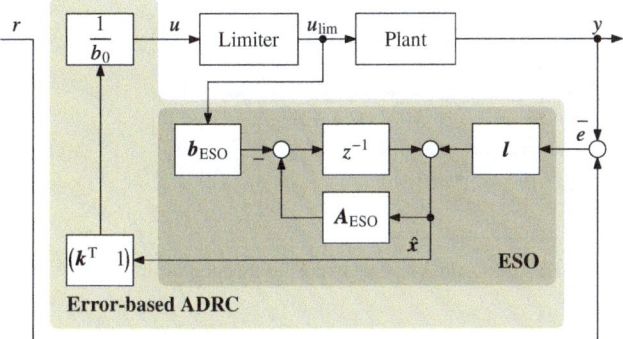

Fig. A.8.1 Discrete-time error-based ADRC, state-space form, with external control signal limiter

A.8.2 Transfer Function Form

Similar to the continuous-time case, the transfer function form of discrete-time error-based ADRC can be obtained from the output-based variant by omitting the prefilter, leaving only a single feedback controller, as shown in Fig. A.8.2:

$$u(z) = C_{\text{FB}}(z) \cdot e(z). \tag{A.8.3}$$

The structure of $C_{\text{FB}}(z)$ and all coefficients are identical to the output-based counterpart in Appendix A.6, i.e., $C_{\text{FB}}(z)$ can be found in Table A.6.1 and its coefficients for the first- and second-order cases in Tables A.6.2 and A.6.3.

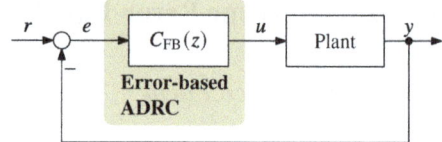

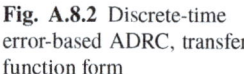
Fig. A.8.2 Discrete-time error-based ADRC, transfer function form

A.8.3 Dual-Feedback Transfer Function Form

A dual-feedback form of discrete-time error-based ADRC reuses the same two transfer functions as in the output-based case from Appendix A.7:

$$u(z) = C_{\text{FBy}}(z) \cdot e(z) + C_{\text{FBu}}(z) \cdot u_{\text{lim}}(z). \quad (A.8.4)$$

Terms $C_{\text{FBy}}(z)$ and $C_{\text{FBu}}(z)$ are given in detail in Table A.7.1 and their coefficients for the first- and second-order cases in Tables A.7.2 and A.7.3. A minimum-footprint implementation is just as well possible and shown in Fig. A.8.4.

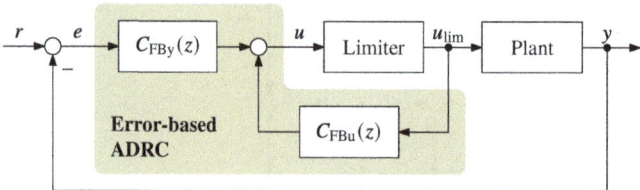

Fig. A.8.3 Discrete-time error-based ADRC, dual-feedback transfer function form

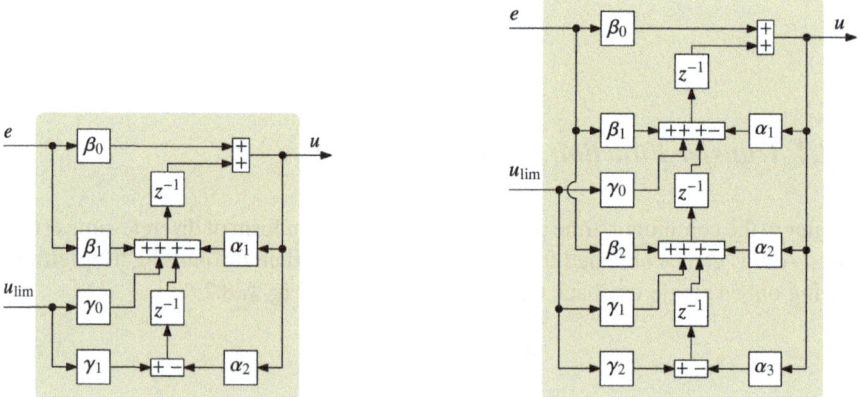

Fig. A.8.4 Minimum-footprint form of first- and second-order discrete-time error-based ADRCs based on the form in Fig. A.8.3. Coefficients are given in Tables A.7.2 and A.7.3

Appendix B
Overview of ADRC Implementations in Simulink

Several model-based implementations of ADRC variants are already available for MATLAB/Simulink, lowering the barrier to evaluate and use ADRC in an application. This chapter briefly describes three different options, covering both commercial and free third-party offerings. The latter include the *Linear ADRC Blockset*, a collection of most of the practically relevant ADRC forms introduced in previous chapters, which is provided as supplementary material to this book. Most importantly, all offerings are classified and linked to the actual control and observer equations provided throughout the book.

B.1 Simulink Control Design: ADRC Block

Starting with MATLAB release R2022b, an ADRC block is offered as part of the *Simulink Control Design* toolbox by Mathworks.[1] Figure B.1.1 shows the block symbol and its possible configuration parameters. This block implements "standard" linear ADRC as understood in this book for model orders $N = 1$ and $N = 2$. It comes with the following features and options:

- *Continuous-time form:* A state-space implementation is being used, with a control law as given in (3.15) and the extended state observer as in (3.16). This form was summarized in Appendix A.1 of this book.
- *Discrete-time form:* This implementation is carried out in state space, as well. The control law is given in (8.1). As in Sect. 8.1 of this book, the observer uses zero-order hold discretization and the *current observer* approach, resulting in observer dynamics described by (8.4).
- *Bandwidth parameterization:* Controller and observer gains can be tuned using the *bandwidth parameterization* approach described in Sects. 3.2.1 and 3.2.2, respectively. Note that, for the controller gains, this ADRC block uses a continuous-time

---

[1] *Mathworks, MATLAB, Simulink,* and *Simulink Control Design* are trademarks or registered trademarks of *The MathWorks, Inc.*

B Overview of ADRC Implementations in Simulink

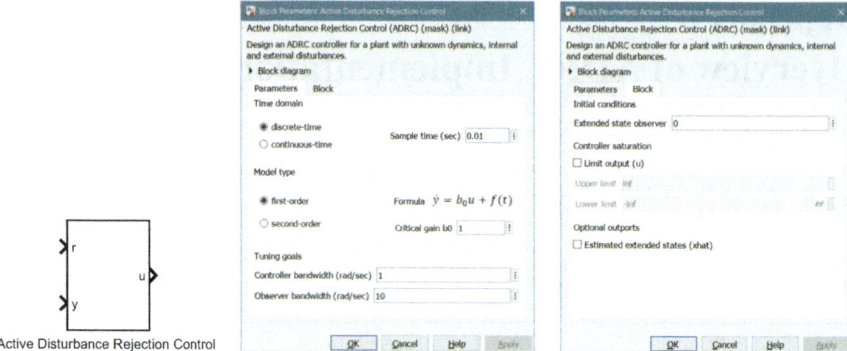

Fig. B.1.1 Symbol and parameters of the *Active Disturbance Rejection Control* block from *Simulink Control Design*

design for both the continuous- and discrete-time cases, i.e., $k_{1/2}$ are always set as given in Table A.1.1. As discussed in Sect. 8.1, this is a reasonable choice as long as the sampling frequency is sufficiently high compared to the desired bandwidth and should not be a problem in any practical design. Discrete-time observer gains $l_{1/2/3}$ are tuned as given in Table A.5.1.

- *Control signal limitation:* An asymmetric magnitude limit of the controller output can be configured. As described in Sect. 9.1, windup protection is provided by feeding the limited control signal back into the observer.

To help users get started more easily with the ADRC block, three model examples are provided: water level control of a water-tank system (see Fig. B.1.2), voltage control of a DC-DC boost converter, and speed control for a brushless DC motor (BLDC). A word of caution from our side: The performance advantage of ADRC presented there in comparison to PID controllers must, at least partially, be attributed to tuning and implementation details of the latter, which offer some room for improvement.

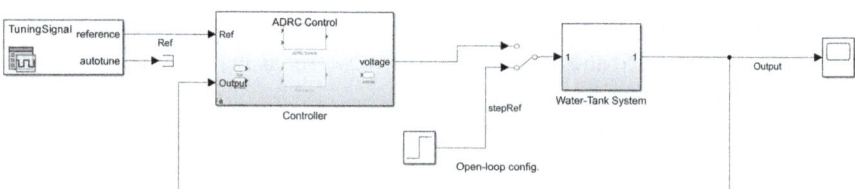

Fig. B.1.2 One of the examples provided with *Simulink Control Design*: control of a water-tank model. For readers coming here directly from Chap. 2, this is a possibility to explore the possibilities of applying ADRC to a nonlinear plant similar to our introductory example from Fig. 2.1. Be aware that this is a simplified scenario with a continuous-time controller missing practical aspects (which we discuss in Chap. 9) such as dead times, measurement noise, and control signal limitations. This means that you will be able to achieve a control performance that can most likely not be reproduced in a real-world setup—regardless of the controller type employed

B.2 ADRC Toolbox for Simulink

An implementation of error-based ADRC has been available since 2021 with the *Active Disturbance Rejection Control (ADRC) Toolbox* for Simulink created by Krzysztof Łakomy et al.[2] Only one controller block is being provided, as shown in Fig. B.2.1, but this can be customized with several parameters and options, which includes feeding parameter values as additional input signals to the block. Its most important characteristics and features are:

- *Continuous-time form:* The controller provided is a continuous-time state-space implementation of error-based ADRC as described in Sect. 6.4.1.
- *Bandwidth parameterization:* Tuning is performed using *bandwidth parameterization* for both controller and observer. As in the ADRC block from Sect. B.1, the observer bandwidth is parameterized as an absolute value (frequency), not using a relative factor to the desired closed-loop bandwidth, as in this book.
- *Control signal limitation and further add-ons:* An asymmetric magnitude limitation of the controller output can be configured. As a solution to bumplessly enabling the controller ("Anti-peaking"), its output can be activated with a delay, to give the observer time to reach a settled state—this is the *tracking* approach discussed in Sect. 9.2.

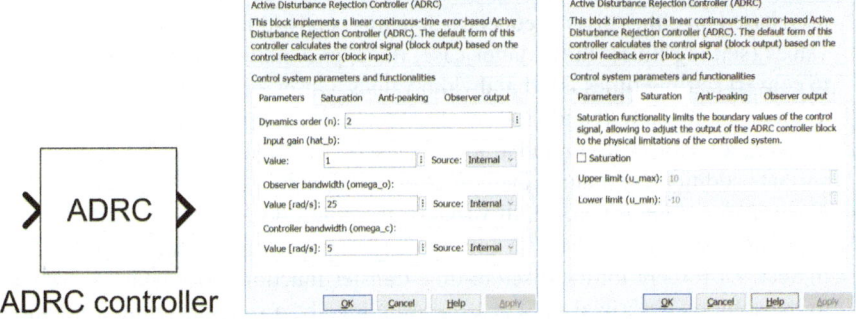

Fig. B.2.1 Symbol and important parameters of the *ADRC controller* block from the *ADRC Toolbox* for Simulink, which provides a state-space implementation of continuous-time error-based ADRC

The *ADRC Toolbox* contains several examples, some of which also allow easy comparison to results obtained with PID controllers: voltage control of a DC-DC buck converter, velocity control of a DC motor, level control of coupled tanks, and temperature control of the TCLab kit described in Sect. 11.1.

---

[2] The *ADRC Toolbox* for Simulink is available for free download from the Mathworks File Exchange at https://www.mathworks.com/matlabcentral/fileexchange/102249. Extensive documentation for this toolbox with several examples is being provided by its authors on arXiv: https://doi.org/10.48550/arXiv.2112.01614.

B.3 Linear ADRC Blockset for Simulink

To get you started with ADRC in Simulink more quickly, we provide an implementation of most of the practically relevant variants discussed here as supplementary material to this book: the *Linear ADRC Blockset* for Simulink.[3] For each of the "classical" (output-based) and error-based ADRC (eADRC) flavors, four variants are being provided, thus covering all options mentioned in the overview, Fig. 12.1, except for the continuous-time transfer function variants, as we do not consider them being primary candidates for deployment in a real-world control system. Characteristics and features of the blockset are:

- *Continuous-time forms:* State-space variants are provided as described in Appendix A.1 (and Appendix A.4 for eADRC).
- *Discrete-time forms:* For output- and error-based ADRCs, three discrete-time variants are provided, respectively:
  - Discrete-time space-space form, as described in Appendix A.5 (and Appendix A.8.1 for eADRC)
  - Discrete-time transfer function form, as described in Appendix A.6 (and Appendix A.8.2 for eADRC), for orders $N = 1$ and $N = 2$
  - Discrete-time dual-feedback transfer function implementation, as described in Appendix A.7 (and Appendix A.8.3 for eADRC), for $N = 1$ and $N = 2$
- *Bandwidth parameterization:* All blocks are tuned using *bandwidth parameterization*, based on either given frequency-domain (bandwidth) or time-domain target values (settling time). In the latter case, the approximation (3.21) is being used to convert settling times into bandwidth values. Observer dynamics can be given as an absolute value (bandwidth, as the blocks from Appendices B.1 and B.2), or using a factor (k_{ESO}) relative to the closed-loop bandwidth. The state-space variants additionally include the *half-gain tuning* option described in Sect. 9.3.
- *Control signal limitation:* In all variants, the controller output (magnitude) can be limited asymmetrically, with anti-windup measured implemented as discussed in Sect. 9.1 (except for the discrete-time transfer function form, which is limited to the solution described in Sect. 8.2). Additionally, a symmetric rate limitation of the controller output can be configured for the discrete-time state-space and dual-feedback transfer function variants.

Figure B.3.1 presents an overview of all blocks provided in this blockset and shows one of the block parameter masks and the implemented block diagram belonging to the latter. As can be seen from the parameter mask, the typical workflow elements for designing and deploying ADRC as described in the cooking recipes of this book can be recognized: plant modeling, controller and observer tuning, implementation, and add-ons (sampling interval and output limitations). The internal realization of these blocks is close to the block diagrams summarized in Appendix A of this book.

---

[3] The *Linear ADRC Blockset* can be downloaded for free from the Mathworks File Exchange at https://www.mathworks.com/matlabcentral/fileexchange/135552.

B Overview of ADRC Implementations in Simulink

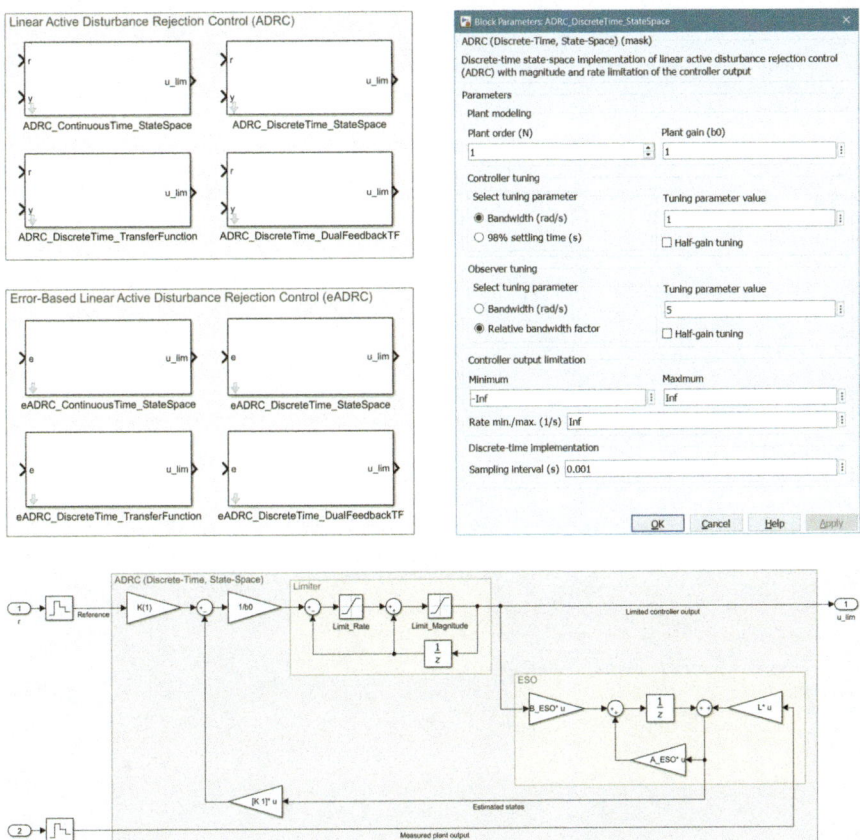

Fig. B.3.1 Overview of the *Linear ADRC Blockset* published as supplementary material to this book, the block parameter mask for one of the blocks (discrete-time state-space variant), and a look into the implementation of the latter

Index

A
ADRC variants
  Continuous time domain
    Non-realizable transfer function form 51, 193
    State space form, 36, 191
    Transfer function form, 52, 195
  Discrete time domain
    Dual-feedback transfer function form, 135, 202
    State space form, 121, 197
    Transfer function form, 131, 199
  Error-based ADRC, *see* Error-based ADRC variants
Anti-windup, *see* Controller output limitation, Magnitude limitation
Application examples
  DC-DC converter voltage control, 177
  Heater temperature control, 171

B
Bandwidth parameterization, *see* Tuning, Bandwidth parameterization
Bumpless transfer, 153

C
Computational footprint
  Comparison, 146
  Minimum-footprint form, 137
Concepts
  Critical gain parameter, *see* Modeling, Plant model, Critical gain parameter
  Disturbance rejection, 31
  Extended state observer, 37
  Integrator chain, 30
  Normalized plant, 31
  Plant gain inversion, 31
  State-feedback controller, 33
  Total disturbance, *see* Modeling, Disturbance model, Total disturbance
Controller output limitation
  Magnitude limitation, 149
  Rate limitation, 151
Cooking recipes, 188

D
Dead time, 159
Discretization
  Current observer, 123
  Zero-order hold discretization, 122

E
Error-based ADRC variants
  Continuous time domain
    State space form, 94, 196
    Transfer function form, 97, 196
  Discrete time domain
    Dual-feedback transfer function form, 141, 206
    State space form, 139, 205
    Transfer function form, 140, 205
Examples
  Application examples, *see* Application examples
  Introductory first-order examples, 11
  Introductory second-order example, 19
  Nominal closed-loop example, 60
Extended state observer, *see* Concepts, Extended state observer

Index

F
Feedback controller, 51, 67
Feedforward, 53

H
Half-gain tuning, *see* Tuning, Half-gain tuning

I
Implementation
  MATLAB/Simulink
    ADRC Block, 207
    ADRC Toolbox, 209
    Custom implementation, 163
    Linear ADRC Blockset, 210
  Source code
    State space form, 165
    Transfer function form, 168
Integrator chain, *see* Concepts, Integrator chain

L
Literature review, 107
Luenberger observer, *see* Observer

M
MATLAB, *see* Implementation, MATLAB/Simulink
Measurement noise, 157
Modeling
  Disturbance model
    Default disturbance model, 47
    Non-constant disturbances, 80
    Total disturbance, 30
  Plant model, 29
    Critical gain parameter, 30, 70
    Parametric uncertainties, 72
    Plant order, 30
    Structural uncertainties, 74
    Using additional plant model information, 79

N
Noise, *see* Measurement noise
Nonlinear ADRC, 84
  Controller, 87
  Observer, 88
  Tracking differentiator, 85

O
Observer, 34, 46

P
Pole placement, *see* Tuning, Pole placement
Prefilter, 51, 99

S
Simulink, *see* Implementation, MATLAB/Simulink
Source code, *see* Implementation, Source code

T
Trajectory generator, 82
Tuning
  Bandwidth parameterization, 39
    Performance analysis, 63
  Comparison of discrete-time and continuous-time tuning, 127
  Controller
    Continuous time domain, 38
    Discrete time domain, 126
    Settling time, 40
  Half-gain tuning, 159
  Observer
    Continuous time domain, 40
    Discrete time domain, 125
    Gang-of-six analysis, 65
  Pole placement, 38

V
Virtual plant, 38

W
Windup protection, *see* Controller output limitation, Magnitude limitation

The manufacturer's authorised representative in the EU is Springer Nature Customer Service Centre GmbH, Europaplatz 3, 69115 Heidelberg, Germany. If you have any concerns regarding our products, please contact ProductSafety@springernature.com

Printed and bound by CPI Group (UK) Ltd, Croydon, CR0 4YY

25/03/2026

02078171-0005